All rights reserved. Copyright © 2023 Aurora S. Santos

Unlocking The Mystery Of Loch Ness

Aurora S. Santos

COPYRIGHT © 2023 Aurora S. Santos

All rights reserved.

No part of this book must be reproduced, stored in a retrieval system, or shared by any means, electronic, mechanical, photocopying, recording, or otherwise, without written permission from the publisher.

Every precaution has been taken in the preparation of this book; still the publisher and author assume no responsibility for errors or omissions. Nor do they assume any liability for damages resulting from the use of the information contained herein.

Legal Notice:

This book is copyright protected and is only meant for your individual use. You are not allowed to amend, distribute, sell, use, quote or paraphrase any of its part without the written consent of the author or publisher.

Introduction

The world of aquatic cryptids is a fascinating realm filled with enigmatic creatures that continue to capture the imagination of enthusiasts and researchers alike. At the heart of this mysterious world lies the infamous Loch Ness Monster, an elusive creature that has fascinated the public for decades. Through detailed accounts and historical evidence, this book sheds light on the enduring legend of the Loch Ness Monster, delving into the various sightings, speculations, and scientific investigations that have shaped its mystique over the years.

As the narrative unfolds, readers are taken on a journey through the extensive timeline of Loch Ness Monster sightings, exploring the compelling stories and controversies that have arisen from eyewitness testimonies and alleged photographic evidence. From the iconic 1970s sightings to present-day investigations, the guide offers an in-depth exploration of the ongoing quest to unravel the truth behind this elusive and enigmatic creature, examining various scientific theories and hypotheses that attempt to demystify the Loch Ness Monster phenomenon.

Beyond the fabled Loch Ness Monster, the guide delves into the intriguing realm of sea serpents, presenting an array of historical accounts and modern-day sightings that highlight the global prevalence of these elusive marine cryptids. Through meticulous research and captivating narratives, readers are introduced to an array of sea serpent legends and their cultural significance across different civilizations, providing a comprehensive understanding of the diverse mythical creatures that have captured the collective imagination of cultures worldwide.

Venturing beyond the Scottish waters, the guide delves into the realm of Canadian lake monsters, shedding light on the mysterious legends of Ogopogo and other cryptids that are believed to inhabit the serene waters of Canadian lakes. Drawing on indigenous folklore and contemporary reports, the guide presents a compelling exploration of these lesser-known aquatic creatures, unveiling the cultural and historical significance that these legends hold within Canadian communities.

Moreover, the guide navigates through the captivating tales of the Lake Champlain Monster and its neighboring cryptids, offering a comprehensive overview of the rich folklore and scientific investigations that have surrounded these elusive creatures. Through detailed analyses and eyewitness accounts, readers gain valuable insights into the diverse range of lake monsters that have become an integral part of the cultural heritage and local traditions in the regions surrounding Lake Champlain.

With a global perspective, the guide expands its exploration to encompass lesser-known American lake monsters, uncovering the intriguing myths and legends that have shaped the folklore of various American communities. Through a meticulous examination of historical records and contemporary sightings, readers are presented with a captivating narrative that highlights the cultural significance and scientific curiosity that these mysterious creatures evoke within the American collective consciousness.

Additionally, this book offers a comprehensive overview of lake monsters worldwide, providing a global perspective on the rich tapestry of aquatic legends and myths that have persisted across different cultures and continents. By showcasing the diverse array of lake monsters and their cultural significance, the guide emphasizes the universal fascination and enduring allure of these mysterious creatures, inviting readers to embark on an explorative journey through the captivating world of aquatic cryptids.

Furthermore, the guide introduces readers to some confounding carcasses that have puzzled scientists and enthusiasts, unraveling the mysteries surrounding the intriguing remains of unknown creatures that have washed ashore or been discovered in various aquatic environments. Through detailed analysis and scientific scrutiny, readers are presented with an in-depth examination of these perplexing carcasses, shedding light on the challenges and complexities that scientists face in identifying and categorizing these enigmatic aquatic remains.

In a compelling exploration of the deep sea, the guide unveils a fascinating array of deep-sea discoveries and monster mimics, highlighting the complex ecosystem and diverse marine life that populate the ocean's depths. From bioluminescent creatures to bizarre deep-sea organisms, readers are introduced to the extraordinary world of deep-sea cryptids and the remarkable adaptations that enable these creatures to thrive in the extreme conditions of the ocean's abyss.

This book serves as an engaging and informative compendium that offers a comprehensive exploration of the captivating world of aquatic cryptids. Through meticulous research, captivating narratives, and scientific scrutiny, the guide invites readers to embark on an immersive journey through the rich folklore, cultural significance, and scientific curiosity that underlie the enduring fascination with these enigmatic creatures of the deep.

Contents

The Loch Ness Monster 1
The Loch Ness Monster – 1970s to Present 25
The Great Unknown: SEA SERPENTS 43
Ogopogo & Other Canadian Lake Monsters... 66
The Lake Champlain Monster & Neighbors 83
Lesser-Known American Lake Monsters 99
Lake Monsters around the World 117
Some Confounding Carcasse 134
Deep Sea Discoveries & Monster Mimics... 152

1

The Loch Ness Monster

"I never believed the monster story, but now must believe the evidence of my own eyes." - NESSIE EYEWITNESS, August 29, 1950

It's not difficult to understand why the Loch Ness Monster, affectionately known as 'Nessie,' has remained one of the most enduring mysteries on our planet. The notion of a great, dragon-like creature lurking beneath the depths of a fathomless lake within Scotland's mist-shrouded Highlands paints a romantic portrait – one that unfailingly stirs the imagination. Nearly a century of controversy, countless hoaxes, rampant tourism, and a slew of scientific denouncements have served to muddy the proverbial waters. Still, the fact remains that hundreds of individuals with otherwise impeccable reputations have sworn that they've observed a massive animal stirring in the loch – something that (scientifically speaking) should simply not be there. For the most part, the descriptions have remained uniform – a huge, dark, smooth-skinned 'hump' or multiple humps breaking the surface of the water and far less frequently having a long, periscope-like neck with small head attached. Moreover, there have been some puzzling photos, films, and sonar contacts through the years that remain unresolved. When combined with a wealth of other data from around the world (to be presented in the following chapters), it suggests an intriguing possibility – that there may indeed exist a species of immense, heretofore undescribed species inhabiting our oceans, as well as certain rivers and lakes of the world, including Loch Ness.

Hippocampus of Greek Mythology

WATER KELPIE LORE

ONE OF THE COMPELLING ASPECTS OF MANY CRYPTIDS (unknown animals) is that there are often longstanding, native traditions that reinforce the modern accounts. This is true of Bigfoot/Sasquatch, the Yeti, and Thunderbirds, as well as our aquatic monsters. With regard to Nessie, Scotland's ancient people firmly believed in shape-shifting water spirits known as Kelpies, which were greatly feared due to their penchant for drowning unsuspecting victims by dragging them down to their watery graves. While like many myths there seem to be various interpretations, the Kelpie legends generally portray them as resembling horses. The etymology of the name, in fact, seems to have been derived from an old Gaelic word meaning horse. Interestingly, there is a similarity to the so-called Hippocampus or 'sea-horse' of Greek mythology. The real significance of this, as you will learn in the following pages, is eyewitnesses of both Nessie and other sea monsters worldwide frequently describe "horse-like" heads sitting atop long necks. Less routinely, observers mention hair resembling a mane, or even vibrissae (whiskers). Now, I'm not for a moment suggesting that there is a species of equine that had adapted to an aquatic environment – simply that there are certain features on these creatures that may superficially resemble those of horses or other hooved animals.

ST. COLUMBA'S WATER BEAST

VIRTUALLY ALL LITERATURE AND TV PROGRAMS ABOUT THE Loch Ness Monster have referenced an incident that allegedly took place in the year 565 AD (in actuality 580 AD), which is intended to establish that encounters with Nessie go back at least that far. The account involves St. Columba of Ireland, who was engaged in missionary work in northern Scotland during that time in an attempt to convert the local people (known as the Picts) to Christianity. According to his biography, which was written a century later, St. Columba was attempting to cross the River Ness (not the loch itself) when he witnessed the burial of a swimmer who had recently succumbed to the bite of an unknown "water beast." When Columba instructed one of his monks to swim across the shallow river in order to secure a raft on the opposite bank, the ferocious creature evidently surfaced once again, roaring and baring its teeth. At that moment, the saint supposedly invoked the name of God and ordered the monster to retreat, which it apparently did. It all certainly makes for a dramatic story. Yet, according to the late historian Charles Thomas, the account should not be taken literally, as religious writings of that time often embellished actual events in order to convey the power and influence of Christianity. In essence, in all likelihood the tale was merely intended to be a parable and may have in reality involved some aggressive, aquatic animal that was unfamiliar to the locals, such as a seal or walrus. Thomas also accurately pointed out that technically St. Columba's purported encounter did not even occur at Loch Ness, but rather several miles away.

SETTING THE STAGE – THE HABITAT

LOCH NESS IS PART OF SCOTLAND'S GREAT GLEN FAULT, A geological rift that was formed about four hundred million years ago during the Devonian period as mighty, continental, tectonic plates clashed. The loch was subsequently filled by melting, glacial water about ten thousand years ago and has remained basically unchanged ever since. Its surface is fifty-two feet above sea level and is relatively long and narrow: 22.5 miles in length and only 1.7 miles wide. However, due to steep inclines on either side, Loch Ness is incredibly deep in parts, with 754 feet being the long-accepted, official greatest depth. Although, in recent years there have been scientifically unverified depth readings of up to 889 feet in spots. The sheer volume of water in the lake is staggering – close to 263 billion cubic feet. It's been said that you could fit the entire population of the Earth within the loch's depths three times over.

The deeper water maintains a constant temperature of around 42°F, so the lake never freezes – and due to peat and other sediments running perpetually into it, Loch Ness' waters are inky and dark. Sunlight, in fact, does not penetrate more than ten feet below the surface. As a result, Loch Ness is considered nutrient poor and is known as an oligotrophic lake, since algae and plankton (the building blocks of a freshwater ecosystem) cannot thrive. Hence, there are only an estimated twenty to thirty tons of fish inhabiting the loch, not counting seasonal salmon runs. Eleven native fish species include eels, trout, Arctic char, pike, and lampreys. Otters are known to be endemic, but are rarely seen. Occasionally, seals from the nearby North Sea enter the loch via the connecting Ness River. Loch Ness is also known for generating various visual anomalies, including surface mirages, floating debris, weird currents, and standing waves called seiches.

Early postcard depicting the Loch Ness Monster (Public Domain)

1933

THE MODERN STORY OF THE LOCH NESS MONSTER BEGAN around 3:00 p.m. on the afternoon of April 14th, 1933. A new road (named A82) that ran along the loch's northern shore had recently been completed, and a substantial amount of old growth had been cleared away. Visibility of the once largely obscured lake was now at an all-time high. Local innkeepers John and Aldie Mackay were motoring along the loch on their way home from the town of Inverness when suddenly, Aldie noticed a "tremendous upheaval" break the mirror-like surface of the water about two hundred yards out. According to popular retellings of the story, the astonished woman shouted to her husband, "Stop! The Beast!" Aldie would later describe what she observed next: something very large moving just beneath the surface, then two, "wet, black" humps rising up and rolling forward, water cascading off – each hummock about six to seven feet in length, with at least six feet in between – for a total visible length of about twenty feet. Aldie's overall impression was of a "whale-like fish," and indeed she had watched many whales along the Scottish coast in her lifetime, although this creature was somehow different. By the time Aldie's husband John had managed to pull the vehicle over and stop, all he observed was a tremendous wash created by something that had just submerged.

Word of the remarkable incident eventually found its way to the Mackays' friend, Alex Campbell, a water bailiff and news correspondent for the *Inverness Courier*. (Incidentally, Campbell would later claim to have numerous sightings of Nessie himself). On May 02nd, the *Courier* published an article about the Mackays' encounter, with the editor insisting on referring to the unknown animal as a "monster." Truth be told, people around the world already had monsters on their minds at that time, since a brand new movie titled *King Kong* was sweeping the globe. Regardless, word of Scotland's strange, new inhabitant spread throughout that year and captured the public's imagination. A modern legend was born.

Other reports from locals soon began to stream in; common folk who had long been reticent to come forward for fear of being ridiculed, or being labeled as delusional or drunk. As Alex Campbell had noted in his article – "Loch Ness has for generations been credited with being the home of a fearsome-looking monster, but, somehow or other, the 'water-kelpie,' as the legendary creature is called, has always been regarded as a myth, if not a joke."

SPICER LAND SIGHTING

THE NEXT MAJOR EVENT OCCURRED A FEW MONTHS AFTER THE Mackay sighting on July 22nd. George Spicer, the director of a prestigious London firm of tailors, was returning home from holiday with his wife around 4:00 p.m. in the afternoon. As the couple motored along the southern edge of Loch Ness, they suddenly noticed something descending an embankment onto the road a couple of hundred yards ahead of them. According to statements written by Mr. Spicer, as well as an interview conducted by skillful Nessie investigator Lieutenant-Commander Rupert T. Gould, the neck of the creature resembled an elephant's trunk, undulating up and down and attached to a "ponderous" body that appeared dirty gray in color and about four feet high. Because they were driving up a slight incline with a rise at the top, the Spicers did not observe the animal's bottom part or any limbs, though they noted an animate structure near the front of the figure, which resembled either a small lamb being carried, or perhaps the end of the monster's tail curling around its torso. Overall, the Spicers felt as though the "abomination" looked "loathsome" and "repulsive" as it crossed the road in a series of jerking motions. Ultimately, the entire affair only lasted a few seconds. The thing disappeared into some brush and then presumably descended another fifty feet downward into the loch.

When the Spicers slowed down and looked where the animal had crossed the road, they noticed a swath of vegetation that had been flattened, where it had evidently plowed through. One troubling aspect of Mr. Spicer's testimony was that through the years, his estimates of the creature's length fluctuated from only six to eight feet, initially, to eventually all the way up to twenty-five feet. Yet, people are often poor judges of size from a distance, particularly when they are excited and seeing something they aren't familiar with. The road had been about twelve feet wide at the time – and the beast's body certainly seemed to span that distance. George sent a letter to the *Inverness Courier* in early August, describing the incident and insisting that both he and his wife would swear an oath if necessary. He claimed that he had no prior knowledge of the Loch Ness Monster before his encounter, which is plausible since it had only been in the local news for a few months at that point and was not widely known of outside of Scotland. Case in point: the Spicer's story wasn't widely circulated until a few months after the incident. It's worth mentioning that alleged sightings of the Loch Ness Monster on land are exceptionally rare.

Hugh Gray's photo of Nessie thrashing about on the surface (Public Domain)

HUGH GRAY PHOTO

THE FIRST PUTATIVE PHOTOGRAPH OF NESSIE WAS TAKEN BY A man named Hugh Gray – a longtime fitter (welder) and resident of the village of Foyers. Hugh had been fond of Sunday morning strolls along the shore near his home and often brought along a camera in order to take snapshots of the enchanting landscape. On November 12th, 1933, the loch was calm, "like a mill pond," when Gray noticed a tremendous disturbance in the water about one hundred yards offshore. He quickly realized that the cause was some type of immense animal, the likes of which he had never seen before. Acting quickly, Hugh snapped five photos of the creature, which he guessed was between twenty to forty feet in length and comprised of what appeared to be a long, dark gray, back attached to a great, lashing tail. Gray never saw the monster's head, which remained submerged under the water the whole time. The entire incident only lasted a few minutes before the thing finally sank down and vanished from sight.

Thinking that the resulting images would probably be inconclusive, Hugh did not take the film to be developed right away. He confessed later that he was also fearful of being ridiculed if that were ultimately to be the case. In the end, Hugh's brother was the one who ended up taking the film in to be processed some three weeks after the incident – and only one of the five exposures revealed anything at all. Still, the resulting photo was published by Scotland's *Daily Record and Mail* newspaper on December 06th and the image, though still open to interpretation after all of these years, does appear to show a twisting, serpentine form that is throwing up a considerable amount of spray. The paper evidently "touched up" the photo for publication – and that fact, as well as general incredulity toward the possibility of Nessie's existence, served to fan the fires of controversy.

Nevertheless, the revelation of that particular image, in conjunction with big London newspapers (which offered wider circulation) publishing George Spicer's account, resulted in popular interest in Nessie peaking by December of 1933. So much was the fervor that circus owner Bertram Mills offered a reward in the sum of twenty thousand pounds for the monster's capture and London's *Daily Mail* newspaper announced it was going to sponsor a proper expedition in order to get to the bottom of the matter. The periodical enlisted the aid of a flamboyant, self-styled "big game hunter" by the name of Marmaduke Wetherell, (in truth an out-of-work, silent film actor), in order to lead the inquiry. Within days of his deployment, Wetherell claimed that he had discovered the spoor of the Loch Ness Monster on the shore – footprints which were quickly discovered to be fakes. Upon examining a cast of the tracks, zoologists at the London Natural History Museum declared that the impressions were identical to those of an immature hippopotamus. In retrospect, it appears that Wetherell had faked the prints using a taxidermy hippo-foot ashtray that he owned. Embarrassed by the fallout, the *Daily Mail* quickly pulled the plug on the expedition.

Portrayal of Arthur Grant's alleged land sighting of Nessie © Bill Rebsamen

ARTHUR GRANT LAND SIGHTING

NESSIE FEVER CARRIED OVER INTO THE NEXT YEAR. NEWS would break that, on January 05th, 1934, a veterinary student named Arthur Grant had been motorcycling home at 1:30 a.m. when he had a sensational encounter with the Loch Ness Monster on the shore. Eerily similar to the Spicer sighting six months prior, Grant claimed that at the time he was on the northern side of the loch (near to the

Mackay sighting location) when a weird, eighteen-foot-long animal bounded across the road in front of him, almost causing him to wreck. He described the creature as having a thick body, with a long neck, small, snake-like head with large eyes situated on top, a lengthy, blunt tail, and four, powerful paddles. Though it was late at night, Grant testified that the moon was exceptionally bright and that visibility had been good, so there was no mistaking what he had seen. Following a face-to-face interview, investigator Rupert T. Gould gave young Arthur the benefit of the doubt with regard to his sincerity, though Gould acknowledged that it was likely that Grant had merely misjudged a wayward seal or large otter. Resident Nessie researcher and author Constance Whyte later vouched for the integrity of both Grant, as well as his family, who she knew well. Through the years there have been rumors implying that Grant's alleged encounter may have been concocted. Indeed, to my own satisfaction, the highly sensational story seems to echo the Spicer affair a bit too much for comfort. And again, with at least three thousand Nessie reports on record, there are fewer than three dozen reputed accounts of the monster on land – and few in recent decades.

The so-called Surgeon's Photo, perhaps the most iconic Nessie image
(Public Domain)

THE SURGEON'S PHOTO

WE MUST ACKNOWLEDGE THAT, THROUGHOUT THE YEARS, A considerable amount of hoaxing has surrounded the Loch Ness phenomenon. One of the greatest revelations of the past thirty years came in 1994, when the so-called 'Surgeon's Photo,' inarguably the image of Nessie that has become most synonymous with the mystery, was revealed to almost certainly be an elaborate hoax. The discovery was the result of two enterprising Nessie investigators, David Martin and Alastair Boyd, who'd come across an obscure 1975 newspaper article alluding to a confession by a man who had boasted how he'd helped to fake, "A famous Nessie photo back in 1934." When the intrepid researchers managed to track down the confessor's brother-in-law, it was none other than Christian Spurling, stepson to the infamous Marmaduke Wetherell – the man who'd been unceremoniously dumped by the *Daily Mail* following his stunt with the hoaxed Nessie tracks. According to Spurling, Wetherell had hatched a plan to get back at the newspaper by serving them up a fake photo. But, realizing that the editors would be suspicious of anything he was associated with, Wetherell recruited a respected London surgeon named Robert Kenneth Wilson to be the front man for his little scheme. Despite the fact that he had a trustworthy middle name, Wilson evidently enjoyed a good practical joke from time to time.

Wetherell commissioned Spurling to construct a miniature, fake monster, which was accomplished by attaching a dinosaur-like head and neck to the top of a toy submarine. The clever gaff was then placed in the shallows of Loch Ness and photographed in such a way that it was impossible to judge its size. Dr. Wilson summarily took the undeveloped reel of film to the town of Inverness in order to be processed, claiming that he'd been sightseeing with his camera when he'd photographed an unusual animal in Loch Ness (on April 01st no less)! Fortuitously, the chemist who developed the film suggested that the better of the two resulting exposures be handed directly over to *The Daily Mail*, who ultimately bought the whole story – hook, line and sinker. The celebrated image made the paper's front page on April 21st, 1934. Over the years, the Surgeon's Photo has been featured in countless books, magazines and TV shows, and for sixty years was widely regarded as the best evidence for Nessie's existence.

In retrospect, an interesting example of foreshadowing may have appeared in Constance Whyte's influential book – *More than a Legend*, published in 1955. With regard to the Surgeon's Photo, Whyte wrote: "Doubts soon began to be cast on its authenticity, probably on account of an indefensible hoax perpetrated on the public and on a well-known newspaper only three months earlier (Wetherell's fake hippo prints). The two events are to this day associated in many people's minds, and classed together as bogus... When pressed to say what kind of hoax this could be, people would explain that some kind of model had been constructed and pulled through the water by a hidden cord. Such contraptions had in fact been made and set floating on the loch." It should be noted that at least a couple of leading cryptozoologists continue to question the veracity of the hoax claim with regard to the Surgeon's Photo and consider that particular claim to be a hoax in itself!

Many other alleged Nessie photos that have been exposed as probable hoaxes through the years can be attributed to the likes of Lachlan Stuart (1951), Peter O'Connor (1960), Frank Searle (1972-1976), Tony 'Doc' Shiels (1977) and George Edwards (2012). The presumed motivations behind these deceptions range from good old publicity and money to perhaps an adolescent desire to fool the public in order to appear clever.

SIR EDWARD MOUNTAIN EXPEDITION

THE FIRST CREDIBLE, WELL-ORGANIZED EXPEDITION TO SEARCH for the Loch Ness Monster was launched on July 13th, 1934. The man behind it was Sir Edward Mountain – a maritime insurance mogul who had been renting a castle a few miles from Loch Ness. Mountain had heard about the reports and was intrigued. He decided that he would employ twenty men and station them at various points around the loch – for nine hours shifts each day – and for a continuous span of one month. (Unfavorable weather ultimately extended the project to five weeks) Each man was armed with a set of binoculars and a Kodak box camera. Mountain's right-hand man was Captain James Fraser, who oversaw the operation and supervised the men. Within the first two weeks, the weather was fine and there were a number of incidents where the "monster watchers" reported spotting large, animate objects. A few long-distance photos were even taken. At the conclusion of the five-week surveillance, the result of the effort was eleven sightings of objects of unknown origin, together with twenty-one nonconclusive photos, mostly showing ambiguous wakes — although five of the best images were at least worthy of discussion and were published. There was also a 16mm telephoto movie clip that had been captured by Captain Fraser following the conclusion of the official expedition on the foggy morning of September 15th.

Fraser's resulting film segment was shown to a handful of scientists who largely concluded the footage merely showed a common seal. That was also the verdict of two zoologists from England's Linnean Society, who viewed the film a few weeks later, although one academic suggested the creature in question moved like a whale. When the scientists' conclusions were published in the prestigious British journal, *Nature,* at least two experienced seal hunters wrote letters to the editor, arguing that seals are exceptionally gregarious, highly visible animals that spend most of their time basking on the shoreline – and that they don't really disturb the water when they swim, nor do they display humps that extend out of the water.

Thereafter, things around Loch Ness seemed to quiet down for a time. The subsequent news cycle that would span from the mid 1930s up until the start of the 1950s was obviously dominated by World War II – and few Nessie encounters managed to make it past the local highlanders. Security around the loch (as well as all major waterways throughout the United Kingdom) was understandably tight during that period. However, according to author Constance Whyte, persistent Nessie sightings were logged during that stretch, including some by military personnel who had been charged with patrolling the loch. One notable encounter, which occurred in April of 1947, involved county clerk J. W. MacKillop and three passengers who were riding with him in his car at the time. All agreed that they had observed "a large, moving object," in the loch – two visible humps with a leading structure like a head that was held out front. The matter was addressed at the next meeting of the Inverness City Council, and according to a subsequent article in the *Inverness*

Courier, still another vehicle carrying four occupants had also claimed to have seen the monster around that very same time from a different vantage point.

TRAGEDY STRIKES
LOCH NESS WAS BACK IN THE NEWS FOR A VERY SAD REASON on September 29th, 1952. Three-time land speed champion John Cobb was attempting to set a new world record for velocity on the water. Cobb had acquired a souped-up, 6,500-horsepower speedboat named *Crusader* and was traveling over 206 miles per hour in a straight line on the loch when his craft suddenly struck a mysterious wave, causing his vessel to literally disintegrate. Tragically, Cobb was killed instantly. The wrecked remains of *Crusader* were only discovered on the bottom of the loch fairly recently, in 2005. As you might expect, some Nessie theorists have speculated that the ominous wake that led to Cobb's death may have been generated by the monster.

FIRST SONAR CONTACT – RIVAL III
FROM MY PERSPECTIVE, ONE OF THE TRULY COMPELLING LINES of evidence with regard to the Loch Ness phenomenon is that, on multiple occasions throughout the years sonar units have recorded large, unknown, moving targets within the loch's depths. Sonar is essentially a technology where sound waves are reflected off of objects beneath the surface of the water and the resulting echoes, which bounce back, indicate the approximate shape and size of various items. The first Nessie echo sounding to gain widespread notoriety occurred around 11:30 a.m. on December 02nd, 1954. A naval drifter (modified fishing boat) named the *Rival III* had entered the loch from the sea and was cruising just east of Urquhart Castle when the onboard sonar equipment detected a sizable object directly below the ship – at a depth of about 480 feet below the

surface and only one hundred feet from the bottom. The resulting readout seemed to show an immense, serpentine shape. Experts who analyzed a paper printout of the event verified that the reading had been genuine and that the contact seemed solid, not indicative of a school of fish or mass of plankton or vegetation. Other examples would follow. During April of 1968, a team led by an electrical engineer named Dr. Denys Tucker from the University of Birmingham, experimented with a pier-mounted sonar unit. According to a paper published in a respected scientific journal, Tucker's team recorded several moving targets indicative of large animals "several meters" in length.

Peter MacNab's photo of Nessie near the ruins of Urquhart Castle (Public Domain)

PETER MACNAB PHOTO

ONE OF THE LAST OF THE CLASSIC NESSIE PHOTOS TO HAVE graced the front pages of newspapers was taken by a vacationing bank director named Peter MacNab. Due to some ribbing that he endured immediately following the incident, Peter kept the photo under wraps for two years before sending a copy (along with a letter) to author Constance Whyte, as well as to the editor of *The Scotsman* newspaper. According to MacNab's testimony, on the late morning of July 29th, 1955, he and his son had stopped their car along the road above Loch Ness, and as the young man attended to something under the hood of the car, Peter got out his camera in order to take a picture of famous Urquhart Castle. Just then, he noticed, "a bulging, long eddy," out in the bay and immediately thought of the monster. Contemplating the distance, MacNab hurriedly attached a telephoto lens to his camera. Then, looking up, he claimed he observed a lengthy, dark shape undulating through the water at a decent clip. Peter was able to snap a couple of shots before the mysterious object submerged. Now, like all reputed photos of the Loch Ness monster, MacNab's image remains controversial. Skeptics have argued that what he saw and photographed was merely a standing wave. Cryptozoologist Roy P. Mackal noted that when he requested a negative of the photo from MacNab for analysis, it displayed some noticeable differences from the original print that had been circulated, raising even more questions. Still, the multi-humped object that's visible in the photo appears to be almost as long as the tower of Urquhart Castle is high – at least forty feet.

DINSDALE FILM
IN THE FOLLOWING DECADES, GENUINE SCIENTIFIC EFFORTS finally began to manifest that would address the Loch Ness puzzle. In 1960, a joint expedition between Oxford and Cambridge Universities strategically deployed stationary cameras around the loch, as well as utilizing sonar equipment. That very same year, an

enthusiastic aeronautical engineer named Tim Dinsdale arrived on the scene. Seduced by the lure of the monster after reading a magazine article in March of 1959, Dinsdale decided to organize his very own one-man expedition – and like Lt. Cmdr. Rupert T. Gould twenty-seven years earlier, he resolved to attack the problem in a highly methodical and scientific manner. First, he spent months undertaking a thorough analysis of all of the eyewitness accounts that had been documented in order to determine the creature's likely appearance and behavior patterns. He then borrowed a high-performance, 16mm Bolex cine camera from zoologist Maurice Burton. Dinsdale removed the passenger seat from his car and mounted a swiveling tripod in its place, and he even rehearsed being in a state of constant readiness – able to film on a moment's notice.

His efforts seemed to have paid off in the morning of the sixth day of his adventure, April 23rd, 1960. (In the course of his research, Dinsdale had noted that 85 percent of Nessie sightings had taken place before 9:30 a.m.). As he drove beside the loch, Dinsdale spotted something unusual on the surface several hundred yards distant, and peering through binoculars, he observed what he later described as a great hump – mahogany in color, though with a noticeable "dark blotch" on one flank. He likened it to the backside of a buffalo. Acting quickly, Tim stopped the car and began filming in rapid machine gun bursts as the object moved with great force toward the opposite shore. Within a span of four minutes, the hump made a sharp-left, ninety-degree turn, its V-shaped wake following the shoreline, after briefly submerging or vanishing from sight. The resulting fifty feet of black and white footage has been carefully analyzed and hotly debated in the sixty years since. Many skeptics have come to the conclusion that what Dinsdale actually filmed that day was a small boat, and that he had been fooled by the poor lighting conditions at the time. To his credit, within hours of the incident, Dinsdale arranged for a fishing skiff (small boat) to travel the same course across the loch that the anomaly had taken while he filmed it from the same spot he'd been at earlier that morning for comparison purposes.

When Tim debuted his footage on a popular British TV show called *Panorama*, Nessie fever struck the world once again. In late 1965, Dinsdale's film was analyzed by England's Joint Air Reconnaissance Intelligence Centre (JARIC), which concluded that the object shown in the film was "likely animate," not a boat – dark and triangular in shape, five and a half feet wide and raised up three feet, seven inches out of the water. JARIC also estimated that the object had been traveling at a speed of ten miles per hour. Recently, through new enhancement methods with modern software, a fortifying case has been made that an overzealous Dinsdale may have in fact misidentified and filmed a skiff that he had simply misjudged from a distance. We must remember that he was at least thirteen hundred yards away from the object when he first spotted it, and investigators have recently learned that his binoculars at the time may have been sub-par. In my humble opinion, Dinsdale's control footage of the boat looks noticeably different from his original sequence, and that was also the conclusion of the experts at JARIC. But the sun was also higher in the sky when the boat was filmed. At any rate, Tim Dinsdale became so affected by the experience that he would spend the next twenty-seven years of his life passionately searching for definitive proof of the Loch Ness Monster.

LOCH NESS INVESTIGATION BUREAU
IF THE TIM DINSDALE FILM FAILED TO PROVIDE CONCLUSIVE proof that a giant, unknown beast inhabited the Scottish lake, it did at least inspire others to become involved in the search. One of those men was David James, a distinguished member of British Parliament and war hero. Initially reluctant to have his reputation sullied by any association with the monster, he was begrudgingly recruited to the cause by famed naturalists Sir Peter Scott and

Richard Fitter – and along with Constance Whyte, formed the Loch Ness Phenomenon Investigation Bureau (LNIB) in 1962. The initial goal of the project was to "Carry on the work of recording and analyzing local eyewitness accounts." However, when James approached a well-connected television executive and expressed his opinion that Nessie might be a nocturnal creature, the mogul responded by donating two intense searchlights (with dedicated power generators) to the cause. This led to the bureau organizing two dozen volunteers who were willing to head up to Loch Ness on a two-week, nighttime vigil during October 1962. Though somewhat disorganized and ultimately unsuccessful on their first attempt, the following year the bureau became more ambitious and grew to some eighty volunteers who would continuously operate ten strategically located observation/photography stations for two weeks during the month of June.

When all was said and done, the LNIB's (the word 'Phenomenon' was ultimately dropped from the title) efforts would span a decade. Each year, dozens of volunteers (mostly students and amateur monster enthusiasts) would return to Loch Ness in order to staff platforms fitted with long-range camera rigs. The group lived caravan-style in a camp along the water's edge. Loch Ness investigator Adrian Shine has likened it all to a type of protest movement – a demonstration against the "arrogance of science."

Until it closed up shop in 1972, the LNIB logged at least 197 incontrovertible sightings, and many team members even reported seeing mysterious humps and, more frequently, substantial wakes created by some unseen forces beneath the surface. Several inconclusive photos and films were also obtained – the most diagnostic being some cine footage that was captured by teenage volunteer Dick Raynor on June 13th, 1967, which shows the linear wash of some unidentified animal moving just under the water. Yet in later years, Raynor concluded that he had merely filmed a line of swimming birds that day.

An illustrious chapter of the LNIB saga began in the fall of 1965, when a vacationing biochemist from the University of Chicago struck up a conversation along the shore with team members who were operating a photography rig mounted to the top of a van. The curious scientist, Dr. Roy P. Mackal, would soon become fast friends with David James and was quickly coaxed into becoming a director for the bureau. Perhaps most significantly, Mackal was able — for the first time in the project's history – to raise meaningful funding through grants from the likes of Chicago's Adventurers' Club and even the publishers of *World Book Encyclopedia*. That money, in conjunction with Mackal's scientific talent, would go a long way in terms of elevating the project to another level. Not only was the bureau able to significantly upgrade their equipment (as well as their strategy for trying to obtain better evidence) but for the first time in Nessie's history, she was not being viewed as merely some kind of joke or publicity stunt. Men and women with impressive academic backgrounds began to join in the hunt.

LONGEST SIGHTING

THE LONGEST CONTINUOUS NESSIE SIGHTING ON RECORD occurred around 10:30 p.m. on the evening of June 15th, 1965. (Keep in mind that the sun sets late at Loch Ness during the summer). The incident involved nine eyewitnesses – including an experienced LNIB veteran, an Inverness police sergeant and the county surveyor. The observation lasted for close to an hour. F. W. "Ted" Holiday was a journalist and angler by profession who had joined the Loch Ness Investigation Bureau in 1963 before deciding to strike out and mount his very own search two years later. After hundreds of hours of studying the loch, Holiday had become intimately familiar with all of its qualities. As he stood watch by Urquhart Castle that particular night, he noticed a distant object in the water that he quickly determined was not a boat. Acting fast, Holiday jumped in his vehicle and sped down the road in order to get closer to the anomaly. He had an excellent telephoto camera on hand at the time and desperately wanted to get a good picture. Stopping at the Clansman Hotel, which overlooks the loch, he was able to see the object better, though it was still too far away for a decent photograph. In his 1968 book, *The Great Orm of Loch Ness,* Holiday described the thing as looking like an upturned boat – about thirty feet long and raised up about five to six feet out of the water. It appeared to be yellowish brown in color. Ted also noticed that there were two men on the opposite side of the loch who were also watching the creature! While all looked on in wonderment, the great hump moved in spurts, going faster, then slower, twisting and turning, occasionally diving and then resurfacing until it finally submerged for good. Sadly, Holiday never got close enough to get in range for a photo. Although the next day he did manage to track down the two other witnesses he'd seen from across the loch: Sgt. Ian Cameron and his friend William Fraser, who both reinforced Holiday's impression of the animal. The two men stated that they had only been fishing from the shore for about ten minutes when they'd first noticed the "whale hump." Cameron later noted that in addition to Holiday there had also been six other witnesses who'd observed the monster from the west side of the loch.

SUBMARINES AND SONAR CONTACT

IN 1969, SUBMARINES WERE FIRST DEPLOYED IN THE SEARCH for Nessie. The trailblazer was a yellow, one-man sub that was built and piloted by an enterprising American named Dan Taylor – in conjunction with the LNIB and sponsored by *World Book Encyclopedia*. Named *Viperfish*, Taylor's vessel arrived at Loch Ness on June 01st and was armed with sonar, a hydrophone (underwater microphone) for communicating with the surface and even two custom harpoon rigs fitted with special arrows. The brainchild of Roy Mackal, these retrievable "biopsy-darts" were designed to extract a Nessie tissue sample from a distance of fifteen feet or less. Due to the difficult currents of the loch, Taylor had to wait a few days in order for calm weather to manifest. When the sub was finally submerged, it immediately plunged nose first into the bottom silt, getting stuck and losing contact with Mackal's team for twenty minutes. *The Viperfish's* subsequent attempts were not much smoother, as the cold, dark, turbulent waters of Loch Ness created all sorts of operational issues.

Next up was the *Pisces*, a six-man job that had been originally been brought to Loch Ness by a movie company in order to discretely tow a life-size, floating model of Nessie. The fabricated monster was designed to be a prop in the film *The Private Life of Sherlock Holmes* (directed by Hollywood legend Billy Wilder). But when Wilder had its buoyant humps removed so that it would sit lower in the water, the model broke free from its tow cable and sunk like a rock straight to the bottom. Remarkably, in 2016 an underwater drone utilizing state-of-the-art sonar located the wreckage of the sunken movie monster on the loch's floor. In any event, liberated from its primary responsibility, the *Pisces* submarine used the opportunity to undertake some field trials and also do a bit of monster hunting. The plan evidently paid off. As Roy Mackal explained in his 1976 book, *The Monsters of Loch Ness,* "During one cruise, the *Pisces* sonar contacted a large moving target about 50 feet above the loch bottom some 200 yards off. *Pisces* moved forward cautiously, reducing the distance by 100 yards. A few feet closer the target began to recede, moving off and rapidly leaving the *Pisces* and its sonar beam. It was a near miss and very disappointing."

2

The Loch Ness Monster – 1970s to Present

"Many a man has been hanged on less evidence than there is for the Loch Ness Monster." - G. K. CHESTERTON

Underwater Photos – Flippers and Gargoyle Heads

DURING THE SUMMER OF 1970, A NEW ACTOR CAME ON the scene – one who would ultimately mold how the Loch Ness phenomenon would be viewed throughout the decade of the 1970s. Boston-born Robert Rines was a modern renaissance man – a renowned patent lawyer, inventor, and scientist with a degree from Massachusetts Institute of Technology (MIT) who also composed opera on the side. Rines was a determined sort of fellow with an analytical mind and big ideas. In 1963, he founded the Academy of Applied Science, an organization dedicated to the pursuit of some of those big ideas. And after attending a lecture given by Nessie investigator Dr. Roy Mackal at his alma mater, Rines (in conjunction with the Loch Ness Investigation Bureau) brought cutting edge, side-scan sonar equipment to Loch Ness. An intensive study was summarily conducted by side-scan pioneer Martin Klein, who concluded that there were indeed large, moving targets in the loch – presumably huge animals.

The following year, Rines would have a life-changing experience. Early in the evening of June 23rd, 1971, Robert and his wife Carol were having coffee with locals, Wing Commander Basil Cary and his wife Winifred, who lived in a cottage just above Urquhart Castle overlooking the loch. Quite unexpectedly, a huge hump rose up out of the water in the middle of Urquhart Bay and began to move, prompting the four frantic individuals to scurry down closer to the shoreline, telescopes in hand. They all watched the object intently for fifteen minutes and agreed that it appeared to be the twenty-foot back of some massive animal with gray-brown, mottled skin. After moving toward the opposite shore for several minutes, the creature turned around and headed back in their direction, then submerged and vanished from sight. This event affected Rines in a truly profound way. He was now a man on a mission.

The next year, the Academy of Applied Science returned with a shiny, new toy – an underwater, stroboscopic camera designed by a man named Harold "Doc" Edgerton. Doc, a professor emeritus from MIT was famous for inventing the strobe light as well as designing underwater camera systems that had been used by Jacques Cousteau. His rig featured two large, tube-like components, one that housed a powerful, fifty-watt strobe light; the other a 16mm camera. The idea was to ambush Nessie underwater by submerging the setup into the chasm of Loch Ness and then programming it to continually flash and snap photos at regular intervals of forty-five seconds. The plan seemed to pay off just after 1:40 a.m. on the morning of August 07th, 1972, when two images, which have since become known as 'The Flipper Photos' were captured. At the time, Edgerton's rig (affectionately known as "Old Faithful") had been suspended thirty-five feet below one boat, while an adjacent vessel deployed a Raytheon sonar unit aimed in the same direction. The echo sounder suddenly began to record a substantial, moving contact, which was later determined to be an object twenty to thirty feet in length. Numerous salmon could also be observed fleeing the target area, as if they were spooked by something.

Rines returned to the US and had the film carefully developed by specialists at Kodak. Within the time frame of when the sonar anomaly had occurred, there seemed to be some unusual images, including two frames that gave Rines the impression of a giant flipper! He sent the photos off to California's Jet Propulsion Laboratory (JPL) in order to be digitally enhanced, and the augmented versions that he got back looked to him like some huge, fin-like appendage from a gigantic sea creature, complete with a stabilizing rib down the center. Based on the approximate distance between the camera and subject (fifteen to twenty feet), the fin was estimated to be at least six feet in length and three feet across, suggesting a truly gargantuan beast. Understandably, these diagnostic photos were presented as irrefutable proof – at long last – that Nessie actually did exist. The pictures were duly published, and famed British Zoologist Sir Peter Scott even proposed a scientific came for the species – *Nessiteras rhombopteryx* (meaning "Diamond-Finned Wonder of Ness"). Surprisingly however, the flipper photos did not have the impact that Rines and other Nessie advocates expected. In the decades since, it's been determined that JPL's "enhancements" may have included overzealous retouching, as the original, unaltered shots are noticeably obtuse in comparison. Or perhaps the first enhancements were enhanced even further by Rines's team. Regardless, independent attempts to recreate the enhancements have not been able to achieve anywhere near the same distinct fin-like shape. Like all reputed images of the Loch Ness Monster, the Flipper Photos remain controversial.

Rines and his group returned to Loch Ness the next two summers with little to show for their efforts. But on June 19th, 1975, lightning seemingly struck twice. A more advanced camera system had been created by Doc Edgerton's protégé, an inventor named

Charles Wyckoff. But Old Faithful was still in use as a backup, although its strobe and camera housings had been distanced in order to reduce glare in the murky water. The tubes were suspended forty-five feet below the surface, while the newer, sonar-activated rig lay on the bottom of the loch, angled upward. At 4:32 a.m., Old Faithful captured a single, color frame of what appeared to be the upper torso, long neck, and small head of some living creature. Since, it was determined to be some twenty-five feet from the camera, the apparition was later estimated to be at least twenty feet in length. A little over seven hours later at 11:45 a.m., something seemingly jostled the camera rig. This is evident because out of seven snapped images, two are of the hull of the boat hovering above. One particular picture was startling and has become famously known as the 'Gargoyle Head Photo.' Though out of focus because the object in frame is only a few feet from the camera lens, it appears to resemble a hideous, reptilian, face, complete with gaping mouth and horn-like structures on top – a veritable gargoyle. Understandably, many Nessie investigators were shocked, since they were unable to relate the image in the photo to the descriptions of the monster that they were familiar with.

In any case, armed with this new evidence, British Parliament member David James scheduled a three-hour symposium before the U.K.'s House of Commons on December 10th, 1975, in order to present a crowning, scientific argument for the existence of the Loch Ness Monster. The standing-room-only audience consisted of hordes of media and also several zoologists. The presenters included Sir Peter Scott, Robert Rines, Doc Edgerton, Martin Klein, Charles Wyckoff, Roy Mackal, and Tim Dinsdale. Water bailiff Alex Campbell, who had first alerted the world of the Loch Ness Monster forty-two years earlier, even recounted his own alleged head and neck sighting from 1934. At the close of the meeting, reactions were mixed. Most academics remained skeptical, feeling that the evidence presented was far too subjective and just not convincing enough to declare that an extraordinary new species existed in a Scottish lake.

One thing the symposium (as well as an accompanying scholarly paper) did accomplish was it generated even more scientific expeditions at Loch Ness during the summer of 1976. Not only did the Academy of Applied Science return once again, but the National Geographic Society joined in the hunt as well. Ultimately, Robert Rines would spend over three decades searching but never managed to obtain any conclusive evidence. As the years went by (perhaps out of frustration), he even speculated that Nessie may have gone extinct.

ALASTAIR BOYD SIGHTING

IN THE FIRST CHAPTER, I MENTIONED INVESTIGATOR ALASTAIR Boyd, who helped to expose the infamous 1934 Surgeon's Photo as being a hoax. However, Boyd was far from being a skeptic of the Loch Ness Monster – in fact quite the opposite – as his interest in the mystery was sparked by his very own sighting! On July 30th, 1979, Alastair and his wife Susan were sitting in their parked car gazing out at the loch, when just over one hundred yards offshore he observed "A large animal… heaving out of the water, something like a whale." Alastair nudged his wife out of the car and the couple raced to the water's edge in order to get a better look. Boyd later estimated that the creature's black, humped back was at least twenty feet long and noted they could both see water cascading off of it as it breached the surface and remained motionless for at least a few seconds. Unfortunately, by the time he and Susan had sprinted back to their car in order to retrieve their camera, the beast had submerged. Although they could both clearly see some huge force just under the surface was creating a massive wake. Boyd later stated, "It was totally extraordinary. It's the most amazing thing I've ever seen in my life." Susan concurred, "Immediately, your mind doesn't want to accept what it's seeing… it keeps searching for some rational explanation." Alastair was so impacted by the event, he spent the next twenty-four years researching Nessie, which included working to expose and eliminate the problematic red herrings.

MORAG

A MERE TWENTY MILES SOUTHWEST OF LOCH NESS LIES LOCH Morar, a similar but smaller Scottish lake, though considerably more remote and much nearer to the sea. Morar also boasts a depth of 1,017 feet in one particular spot, making it one of the deepest bodies of water in Europe. Apparently, the loch is also home to its very own version of Nessie, though the monster is known locally by an old Gaelic name – Morag. In the tradition of the water kelpies, references to this legendary beast date back many years. And in modern times, numerous residents have reported seeing a large object that resembles an "upturned boat" surface in the lake and then move around, as if self-propelled. A sensational encounter that occurred just after 9:00 p.m. on the evening of August 16th, 1969 made newspaper headlines worldwide and momentarily thrust Nessie's cousin Morag into the spotlight.

After a long day of fishing, Duncan McDonnell and William Simpson were heading back to the dock in their motorboat when they heard a disturbance just behind them. Looking back, the men were flabbergasted to see the back of some immense creature approaching rapidly in their wake. Within moments, the animal was astern of their vessel and actually grazed its side, understandably causing the men to panic. While Duncan abandoned the wheel and grabbed an oar with which to fend off their assailant, William picked up his rifle and fired at the thing, causing it to sink down out of sight. Both men testified that the visible back of the monster had been some twenty-five to thirty feet long and that it displayed three low humps. Its skin had looked rough and of a dirty brown color.

Inspired by this particular incident, a segment of the Loch Ness Investigation Bureau decided to move over to Loch Morar in order to launch a separate expedition known as the Loch Morar Survey, which would ultimately span a period of two years. About thirty Morag reports in total were collected by the team, although an array of shoreline camera rigs that were deployed and manned for countless hours failed to produce any evidence. However, two team members, Ronald Binns and Ian Johnson, claimed that they were shadowed by a strange, linear wake while on a midnight vigil in their inflatable raft. This happened toward the end of the two-year survey, on August 10th, 1971.

Following four decades of relative silence, Morag was seen and photographed by holidaymakers Doug and Catherine Christie in August of 2013. At the time, the retired couple were staying at an inn near the lakeshore and claimed on three separate occasions they observed a solid, black shape rise up out of the water for as much as ten minutes at a time. A picture they snapped of the object seems to show a long, dark bulge extending upward and the hint of a massive body just below the surface. The Christies estimated the thing had been at least twenty feet long. Doug later told the *Daily Record,* "I could not believe my eyes. I am not the type to get excited unduly. But, this just couldn't be explained." Upon further reflection, Charlotte added, "I thought it was a whale."

OPERATION DEEPSCAN

IN 1973, A BEWHISKERED, TWENTY-FOUR-YEAR-OLD investigator named Adrian Shine showed up at Loch Morar, hoping to join up with the Loch Morar Survey. Adrian was disappointed to learn the project had already been disbanded. Still, he was enthusiastic about the prospect of an "exciting wildlife mystery" to be pursued and decided to undertake his very own expedition, which he named the Loch Morar Project. Shine noted the water clarity in Morar was greatly superior to that of Loch Ness, resulting in much better visibility. So presumably he would have a better chance of solving the Morag mystery, as opposed to proving the existence of Nessie. Adrian set about constructing a bathyscope, which is basically an airtight, underwater viewing chamber, within which he could be lowered into the depths in order to observe and hopefully film his quarry. He also decided to use video cameras, because he would have an immediate record of both the animal and its natural movements. Yet ultimately, at the conclusion of the 1970s and with no evidence of Morag having been obtained, Shine moved over to Loch Ness in order to begin planning a strategy for thoroughly examining that lake's murky and unexplored depths. This would hopefully be achieved with the aid of improving sonar technology. Thus began something that Shine called the Loch Ness Project.

Adrian's primary vision was actualized during an event called Operation Deepscan, which took place on October 09th, 1987. A flotilla of twenty-five boats, each armed with a Lowrance X-16 sonar unit would scan the loch (nineteen boats cruising side by side in a line, as well as six support boats trailing behind) from one end to the other – a twelve-hour process. The idea was to create a huge, moving, sonic "curtain" that would detect any animal larger than a salmon. The affair turned into a huge press event, with 326 reporters representing twenty-three media outlets from around the globe in attendance. In his own mind, Shine was, "Sweeping the loch clean for science." Over the course of thirteen years, Adrian had become skeptical about the reality of either Morag or Nessie – at least the possibility of the creatures representing an unknown species, or prehistoric survivor. Shine has acknowledged some large fish or seals might be involved.

On the first day, three large, unidentified targets were detected across from Urquhart Castle – from depths ranging from 256 feet to 590 feet below the surface. When an expert from Lowrance examined the data, he stated the contacts appeared to be animals larger than sharks but smaller than whales. This understandably created a great deal of excitement among everyone that evening, especially the press. The following day, however, the experiment was repeated and no contacts whatsoever were made. Seemingly deflated, everyone basically shrugged their shoulders and went home. It must be recognized that in reality only about 60 percent of the loch had been searched by Operation Deepscan, as the boats were unable to navigate many of the bays and inlets. Adrian Shine has since become a recognizable fixture on the Nessie scene, appearing on numerous TV documentaries about the monster. He currently runs the Loch Ness Centre and Exhibition in Drumnadrochit.

Ruins of Urquhart Castle overlooking Loch Ness. Photo Courtesy © Leanne Geraerts

1990S

IF THE INTENT OF OPERATION DEEPSCAN WAS TO SLAY THE proverbial dragon of Loch Ness, it failed to do so. Sightings and alleged photos of Nessie have continued over the past thirty-three years at an average clip of about ten reports annually. Case in point: on August 10th, 1993 a forestry worker named Roland O'Brien was fishing from the shore near the village of Dores when he noticed "A large, dark hump about five hundred yards out." O'Brien estimated the odd bulge was between eight to ten feet in length and extended four feet out of the water at its highest point. He watched the object for several minutes, as it moved rapidly, "making lots of splashing." The thing stopped abruptly a couple of times, sank down, then rose up again, then turned and moved directly toward the astonished witness before finally submerging for good. When the creature turned, O'Brien got the impression its underside was of a lighter color. Roland, who had been fishing the loch for a decade, became incredulous when asked if he might have been mistaken about what he'd seen. He had been understandably reluctant to talk about his sighting after it first happened, fearing he wouldn't be believed.

A year earlier, the first major environmental and biological survey of Loch Ness had begun. Organized by journalist and Nessie investigator Nicholas Witchell and dubbed Project Urquhart, the effort included teams of British scientists in conjunction with the Swedish sonar company Simrad and even the Discovery Channel television network. However, the project's participants were intentionally ignoring anything having to do with the monster. Simrad began by undertaking what is known as a hydrographic survey, using state-of-the-art sonar and some seven million echo soundings in order to intricately map the loch's underwater geological features. Next, biologists came in and extracted forty-one core samples of the microorganisms in the lake – phytoplankton (plant-based) and zooplankton (animal-based), so they could get a sense of the food chain that was present. In addition, extensive trawling was utilized as a way of evaluating the resident fish population. Yet, something curious did happen. At around 7:00 p.m. on July 28th, 1992, the research vessel *Calanus* tracked a sizable, unknown sonar contact for a period of two minutes. The target was substantially larger than the fish traces the team had gotten used to recording. In the words of one operator, "Far too large." By the time the survey concluded the next year, four more large, unidentified objects had been recorded.

MITCHISON VIDEO

ON SEPTEMBER 05TH, 1998, A TOURIST NAMED GEOFF Mitchison was a passenger on the Loch Ness cruise boat, *Nessie Hunter.* As he aimed his video camera out the window, Mitchison noticed something sticking out of the water, parallel to the shoreline near Urquhart Castle. Geoff slowly zoomed in and managed to videotape the peculiar object for just over ten seconds, until it submerged and was lost from sight. At first glance, the resulting footage appears to show the football-shaped cranium of some animal. There even seems to be a dark eye visible. Critics have pointed out the creature's head curiously resembles that of a seal. Yet, the specimen appears to be swimming with its snout held underwater, which a seal would never do. Additionally, the bizarre beast sinks straight down rather than diving forward and revealing its body, also uncharacteristic of a member from the phocid family. Another intriguing detail is that a few birds that fly through the frame look relatively tiny in comparison to the mystery animal. However, admittedly it's difficult to estimate the creature's actual size. The three-year period from 1996 to 1998 was surprisingly active for Nessie sightings, with forty-three reports being documented. In one instance on the afternoon of December 31st, 1998, a local couple observed a great, black hump break the surface for ten to fifteen seconds.

2003 – A SONAR ODYSSEY

THE BRITISH BROADCASTING CORPORATION (BBC) LAUNCHED an expedition during the summer of 2003 for a documentary titled *Searching for the Loch Ness Monster.* A team led by "subsea" logistics expert Ian Florence scanned the loch with six hundred different sonar beams. "We went from shoreline to shoreline, top to bottom on this one, we have covered everything in this loch and we saw no signs of any large living animal in the loch," Florence stated. A psychological experiment was also conducted, whereby a wooden pole was anchored in the loch and allowed to bob up and down in front of a group of unsuspecting tourists, in order to see how many of them would think it was Nessie. When later asked to describe what they had observed, various answers included: "An underwater machine," "submarine" and (tongue-in-cheek) "Nessie's breathing pipe." Evidently, the only person the researchers had temporarily fooled had been an enthusiastic young boy who'd spotted the object from the road above and run down to the shore in order to get a better look. Nevertheless, the BBC gave themselves a big pat on the back and declared they had definitively proven, once and for all, that the Loch Ness Monster did not exist.

Nessie apparently did not get the memo, since on June the 01st of that very same year there were three different sightings within an eight-hour span. At 2:00 p.m., the skipper of the tour boat *Royal Scot* observed the wake of a fast-moving, submerged object beneath the surface of the glassy loch, traveling at speeds over thirty miles per hour. On another excursion six hours later, all twenty-five passengers (and three crew members) onboard watched a mysterious five-foot hump break the surface for close to thirty-five minutes. Both sightings occurred on the south end of the loch, near the village of Fort Augustus. Later that same evening, a fisherman at nearby Borlum Bay also claimed to have seen the mysterious, dark hump moving about.

GORDON HOLMES VIDEO
ONE OF THE MOST COMPELLING PIECES OF FILM EVIDENCE TO date was obtained by Nessie researcher Gordon Holmes at approximately 9:50 p.m. on the evening of May 26th, 2007. The footage was captured with a handheld, consumer-grade video camera, but clearly shows the serpentine shape of some sizable animal slithering just beneath the surface of the water – swimming against a strong current. Skeptic Dick Raynor has argued the disturbance is merely an unusual wave effect, caused by a freak whirlwind. However, in 2013, the entire two-minute, thirty-six second segment was stabilized, enhanced, and analyzed by an American computer wizard named Bill Appleton, who concluded the footage shows a monster-sized "eel" between ten to fifteen feet in length. While Loch Ness is known to be inhabited by an abundance of eels, the largest known endemic specimens are only a few feet long – and the world record for any eel species scarcely approaches ten feet. Despite this reality, the giant eel theory has gained traction among many speculators in recent years and will be discussed at length in Chapter Nine. A population of mutant, monster eels would certainly be an amazing prospect, but doesn't in any way correspond with the vast majority of eyewitness descriptions of Nessie.

ATKINSON SONAR CONTACT
CRUISE BOAT CAPTAIN MARCUS ATKINSON RECORDED A notable sonar contact in the loch on August 24th, 2011. Atkinson had just dropped off some passengers at Urquhart Castle and was steering clear of some other tour boats when he happened to glance over at his sonar unit and noticed a significant target was materializing on the screen. In his many years of working on Loch Ness, the skipper had never seen anything like it. The echo sounding was huge – much larger than any of the countless fish he'd encountered through the years – and it displayed a distinct serpentine shape. By all estimates, the object was solid and almost

five feet wide. Atkinson nimbly grabbed his smart phone and snapped a photo of the monitor screen, saving it for posterity. Before too long, the image found its way into the hands of a few Nessie investigators, as well as some media outlets. For his efforts, Atkinson was recognized for capturing the best Loch Ness Monster photo of 2011 and received a small prize from a publishing company as part of a publicity campaign. At least one academic has suggested Atkinson's sonar reading shows nothing more than a biomass of algae or zooplankton that was riding on top of something called a thermocline, which is a boundary between water layers of dramatically different temperature. Yet, neither Atkinson nor any of the other experienced boat captains on the loch ever remembered encountering such a phenomenon before.

APPLE MAPS PHOTO

IN MID-APRIL 2014, THE INTERNET WAS ABUZZ WITH THE NEWS that Nessie may have been accidentally captured on a satellite image of Loch Ness, which was being displayed in the computer program Apple Maps! The anomaly was brought to the attention of Gary Campbell, a Nessie researcher who heads up The Official Loch Ness Monster Fan Club, and by no less than two different individuals who'd noticed the curious object while looking at the brand new satellite view of the loch. Examining the photo, it was readily apparent something immense (later estimated to be about one hundred feet in length) was creating a massive wake in the loch. Displaying a blunt, U-shape in front and a wide, V-shaped wash extending behind, the image gave one the impression of a monstrous catfish, complete with pectoral fins and a tail. While the disturbance in the water was quite evident, whatever was causing it remained ambiguous, since the culprit seemed to be submerged just below the surface. It wasn't long before an explanation was offered – the object appeared to be the popular tour boat *Jacobite Queen*, which regularly travels across the loch. According to experts knowledgeable about satellite imagery, the process by which various photos were stitched together caused the vessel to essentially blend into the surrounding water, making it seem invisible to the naked eye, although its wake was still plainly visible.

IAN BREMNER PHOTO
WITHOUT QUESTION, THE WORLDWIDE EXPLOSION OF THE
Internet in recent decades has dramatically altered both how we share information and also how we view our world. This holds especially true with regard to provocative subjects like Nessie. For example, in the early part of this century, the first live-stream webcams at Loch Ness were placed at the water's edge, so that anyone could try and have a Nessie sighting from the comfort of their own home. There are still various websites that continue to offer this opportunity. So far, the results have (like everything related to the mystery), been frustratingly controversial. What's more, alleged photos of Nessie are now digital – and can shoot around the globe at lightning speed via social media and different online news feeds, often becoming "viral." A fairly recent example is the picture taken by a fifty-eight-year-old whiskey factory (try not to smirk)

worker named Ian Bremner. On September 09th, 2016, Bremner was out photographing wildlife and took a few, random shots of the loch. It wasn't until later that Ian realized he had evidently captured the image of some mystifying animal, looping through the water – with a head, neck and two, smooth, wet humps clearly on display. Bremner estimated the unknown beast could be no less than two meters in length. Although he'd always considered himself a Nessie skeptic, Ian admitted he might have unwittingly procured convincing evidence of the monster's existence. Alas, the head that's visible in the picture is almost certainly that of a harbor seal – and the two humps trailing behind look unmistakably like the curled backs of two other diving seals following behind in a line.

CHARLOTTE ROBINSON PHOTO

A TWELVE-YEAR-OLD ENGLISH GIRL NAMED CHARLOTTE Robinson was on holiday with her parents during the weekend of August 17th, 2018. Her family was staying at a caravan park in Invermoriston, near the southern end of Loch Ness. At around 7:00 p.m. in the evening, Charlotte went out to take a look across the water since she was fascinated by tales of the monster. Stunningly, just then, a substantial, black object popped up out of the surf about fifty feet in front of her. The quick-thinking Charlotte managed to snap a photo on her iPhone 7. Her first impression was the thing had a "hook" shape to it. After about a minute, the apparent creature submerged, only to reappear briefly in a different location some seven minutes later. When the teen dashed back in order to tell her parents she'd just seen the Loch Ness Monster, they immediately assumed she was simply pulling their leg. Until, that is, they got a look at her picture! Within days, the family paid a visit to longtime Nessie researcher Steve Feltham who instantly declared it was the best Nessie photo taken in years. Several UK newspapers ended up publishing Charlotte's snapshot, which does seem to show the small head and long neck of some sizable animal poking out of the water. Feltham has suggested the creature in the picture somewhat resembles a seal, though he also feels the photo warrants further investigation.

EDNA STUDY

BEGINNING IN 2018, GENETICIST NEIL GEMMELL OF NEW Zealand's University of Otago began an ambitious project that planned to utilize a relatively new methodology known as eDNA in order to determine if the Loch Ness Monster might actually exist. 'Environmental DNA' is a process whereby samples are collected from air, soil, or water and then any biological material is extracted via a filtering technique. Strands of deoxyribonucleic acid can then be isolated and sequenced. Those sequences are then compared to hundreds of millions of samples logged in genetic databases, which allows for identification of various species that have been recently active in that particular environment. Keep in mind all life forms are continually introducing cells into their surroundings, whether from their droppings, saliva, blood, and other excretions, or even the constant shedding of fur or skin particles. Gemmell admitted he was open-minded about the reality of Nessie and felt it would be a useful exercise for the blossoming science.

Over a period of several months, two hundred fifty water samples, each about one liter in volume, were collected from various points and depths around Loch Ness. The tedious process of extraction and sequencing occurred back at the University of Otago and ultimately some three thousand different species of plants and animals were detected, though only fifty-five of those were vertebrates: twenty-two bird types, nineteen mammals, eleven fish and a few amphibians. Because no reptile DNA was found, Gemmell felt it was important to state the study refuted the possibility of a prehistoric, marine reptile such as a plesiosaur – a favorite candidate of many Nessie investigators through the years. Additionally, no trace of either sturgeon, sharks, or giant catfish were found, presumably eliminating those popular Nessie candidates from contention, as well. Curiously, no DNA of either otters or seals were present either, despite the fact that those aquatic mammals are known to visit the loch. Gemmell has emphasized his team was surprised by the abundance of eel DNA present in virtually every single sample and concluded the possibility of giant eels living in the loch couldn't be dismissed. The scientist also admitted that around 25 percent of the DNA collected could not be unidentified, which is evidently fairly standard for any eDNA study. Nessie supporters have quite obviously focused on this particular detail, as well as the fact that DNA is only truly viable in water for a period of about seven to twenty-one days – leaving open the possibility that the Loch Ness Monster migrates out to sea on occasion.

The bottom line is that, like Operation Deepscan and the Surgeon's Photo hoax revelation, the media continually seizes on these events so they are able to tout "The Loch Ness Monster mystery has finally been solved!" Ah… if only life were so simple. Ergo, read on as we continue on our voyage for more answers.

BREAKING NEWS – NEW SONAR CONTACT!

JUST AS I AM FINISHING UP THIS MANUSCRIPT THERE IS exciting, breaking news in the Nessie-sphere. At around 4:30 p.m. on Wednesday, September 30th, Mr. Ronald MacKenzie, skipper of the *Spirit of Loch Ness* tour boat, recorded a remarkable sonar contact some 558 feet deep, just off the bottom of the loch. MacKenzie, who has never before seen anything like it during his thirty-plus years on Loch Ness, has estimated the crescent-shaped object was close to thirty feet in length. The incident only lasted ten seconds or so. Upon docking, Ronald immediately called his friend, thirty-year Nessie investigator Steve Feltham, who has since voiced his opinion that this may be some of the very best evidence ever produced. Steve recently wrote me, "I have never seen such an impressive mid-water contact as that. It's difficult for sceptics to argue with something so clear. It obviously doesn't tell us what it is. But it does clearly show one solid object about 500' down in the middle of the loch. To any doubters of this mystery I would say, 'Well, what is that, then?'"

❦ 3 ❦

The Great Unknown: SEA SERPENTS

Sea Serpent attacking ship as portrayed by Olaus Magnus (Public Domain)

Norway's Fabled Soe Orm

SWEDISH HISTORIAN, THEOLOGIAN, AND cartographer Olaus Magnus first mentioned a dreadful, Norwegian sea monster as early as 1555 AD, when he published his expansive volume on the history and folklore of

Scandinavia, titled *Historia de Gentibus Sepententionalibus (A Description of the Northern Peoples).* In his opus, Magnus described a snake-like leviathan, "of an astonishing size," that grew "upwards of 200 feet," in length and which lived in the fjords along Norway's extensive coastline. According to Magnus, the highly dangerous beast, known as the 'Soe Orm,' (Sea Worm) dwelt mostly in underwater caverns, but would surface in calm weather in order to devour various types of animals – and occasionally even helpless men, whom it hungrily snatched off of the decks of their seaborne vessels.

Nearly two centuries later the Bishop of Bergen, a prelate named Erik Pontoppidan, chronicled even more accounts of the immense, serpentine animal that was said to frequent the fathomless fjords of Norway's western coast. In one particular instance, Captain Lorenz de Ferry of the Royal Norwegian Navy swore an oath before a magistrate with regard to his encounter off the port of Molde in August of 1747. At the time, de Ferry was commanding a small craft being rowed by eight sailors under his command. The colossal creature the crew observed was said to have the general form of a coiled serpent, but possessing of a horse-like head (complete with flowing, milky, white mane), which it held two feet above the surface of the water. De Ferry also testified the animal's eyes and gaping mouth were both pitch black. He had ordered his reluctant crew to draw close to the sea monster. But, realizing they could not keep pace, de Ferry fired upon the animal, at which point it submerged and vanished from sight.

Interestingly enough, there would be other serpent sightings off the coast of Norway, near Molde, as well as the port city of Kristiansand nearly a century later, during the summer of 1846. The accounts were strikingly similar, describing a creature of great length, in the neighborhood of forty to one hundred feet long – definitively snake-like and moving through the water with vertical undulations similar to that of a caterpillar. All were generally in agreement that the animal's unusual mode of locomotion resulted in a series of humps that would bulge out of the water. The beast was usually portrayed as having smooth skin and as being dark brown in color on top, but displaying a yellowish-white underside.

Sea serpent alleged by Hans Egede off the coast of Greenland in 1734
(Public Domain)

EGEDE'S SERPENT

NORWEGIAN-BORN HANS EGEDE WAS KNOWN AS THE "APOSTLE of Greenland," due to the fact that he established missionary colonies while spreading Christianity along the coast of that vast, frozen super-island. On the 06th of July, 1734, while sailing off the coast of the port village of Godthaab, Egede and his crew observed a "terrible monster" breach the surface and, by all accounts, rise up as high as the ship's masthead! As later recounted in his 1740 biography, the titanic creature displayed the general form of a serpent, but with a long snout, as well as two enormous fore-paddles. Egede noted the imposing beast spouted like a whale when it breached, and its skin looked mottled and "uneven." Because Hans Egede was known as an accurate chronicler of historical events and natural history who was intimately familiar with whales and other marine species of the region, it seems unlikely he and his crew encountered an animal that was familiar to them. In fact, this particular incident is considered to be one of the earliest credible accounts of a sea serpent, due to Egede's reputation. Although admittedly, the size he described would infer a beast of truly, mind-boggling proportions. It's noteworthy that the overall impression is of a monstrous mammal, not a reptile, as there were no scales visible and only whales are known to spout.

Whimsical depiction of Massachusetts' Gloucester Harbor Sea Serpent
(Public Domain)

NEW ENGLAND SEA SERPENT

TRAVELING ACROSS THE NORTH ATLANTIC, WE ARRIVE AT THE remarkable events that unfolded during the month of August 1817. No less than dozens (perhaps even hundreds) of respected citizens from the state of Massachusetts claimed to have observed a sea serpent that was disporting itself in Gloucester Bay on a regular basis. The episode in truth created such a stir that a scientific organization known as the Linnaean Society set out to undertake a thorough investigation of the matter, which resulted in eleven sworn eyewitness affidavits testifying to the veracity of the inexplicable events.

In her informative book, *The Great New England Sea Serpent*, author J.P. O'Neill points out the creature had evidently been familiar to the native Pawtucket tribe of that region at least two centuries earlier. And as early as 1639, an explorer named John Josselyn wrote of incident where a sea serpent had been seen, "Coiled up… upon a rock at Cape Ann." When two Englishmen on a passing vessel attempted to fire upon the thing, their Native American companions forbade them to shoot, stating they would surely find themselves in danger of retaliation if they were to miss the monster.

While the first documented sighting from the 1817 flap occurred on August 06th, things really got rolling a few days later on the 10th, when the animal was spotted at close range by a man named William Row near a place called Rocky Neck Cove. According to Row, the monster was around one hundred feet in length and displayed a head that was "as broad as a horse's." The beast moved rapidly through the water and seemingly appeared to be feeding on fish. Then on August 14th, there were two more dramatic encounters. Row's wife Mary had an early morning sighting from their home as she watched the serpent travel between Rocky Neck and Ten Pound Island at a distance. Like her husband, Mary felt the creature was around one hundred feet in length, had a horse-like head that it kept above the water, and that overall it had moved at a fast pace. That very same day between 4:00 p.m. and 5:00 p.m., ship's carpenter Matthew Gaffney was piloting a craft with two passengers onboard when the thing passed within thirty feet of their position. Gaffney, who was armed at the time, fired a musket ball at the beast's head, which caused the animal to turn and submerge, resurfacing about one hundred yards distant. The eyewitness noted the sea serpent's undulations were "vertical like a caterpillar." Gaffney also swore in his affidavit that the barrel-thick serpent had smooth, dark-colored skin on top, with a white belly underneath.

A correspondence published in a Boston newspaper over three decades later seemed to affirm the events of the time. According to its author, a Colonel T. H. Perkins, he had himself traveled from Boston to the Cape on August 18th in order to observe the animal for himself. As he sat on a rise near the shore with a companion and some other hopeful monster spotters, Perkins got his wish. "I saw on the opposite side of the harbor – the snake… moving with a rapid motion… As he approached us, it was easy to see his motion was not that of a common snake… but evidently the vertical movement of the caterpillar. As nearly as I could judge, there was visible at a time about 40 feet of his body… It was evident, however, that his length must be greater than what appeared, as, in his movement, he left a considerable wake in his rear." Through his fine "glass" (telescope), Perkins determined the serpent was of a chocolate color. His letter concluded, "I left the place fully satisfied that the reports in circulation… were essentially correct."

One gaffe that cast a spurious light on the whole 1817 affair occurred when two boys discovered a three-foot, common black snake with a deformed (lumpy) spine at a place called Loblolly Beach. The Linnean Society promptly declared the specimen must be a juvenile sea serpent and even proposed a scientific name – *Scoliophis atlanticus*. It wasn't long before the mistake was exposed and ultimately the entire New England Sea Serpent affair was written off by the scientific community, despite the fact that so many corroborating accounts had been sworn to under oath.

Depiction of the Sea Serpent sighted by the crew of the HMS Daedalus
(Public Domain)

HMS DAEDALUS SIGHTING

WITHOUT QUESTION, THE MOST CELEBRATED SEA SERPENT sighting involves the *HMS* (Her Majesty's Ship) *Daedalus*, a British corvette, which was sailing off of the coast of Southwest Africa on August 06th, 1848. The mysterious creature in question was observed for a twenty-minute span by the vessel's captain, Peter M'Quhae, as well as six other crew members, including Lieutenant Edgar Drummond, Master William Barrett, midshipman Mr. Sartoris, as well as the quartermaster, boatswain's mate, and even the man at the wheel. Being that it was Sunday, the rest of the crew was either having supper or resting below deck at the time. According to both M'Quhae's captain's log as well as Drummond's journal entry, the weather that day was cloudy and "squally" with swells, a stark contrast from the smooth, glassy water conditions that typically produce sea serpent accounts.

At around 5:00 p.m., Sartoris first spotted the subject and immediately alerted the officer of the watch, Lt. Drummond, who was walking the deck with both M'Quhae and Barrett at the time. As all looked on in disbelief, the curious animal swam in a southwesterly direction, across their stern and through the ship's wake with a determined purpose and at a speed approaching thirteen knots. For at least five minutes, the monster was easily in sight – and for the following fifteen minutes, within range of their telescopes. Generally, all were in agreement that the elongate beast was showing about sixty feet of its straight back above the water, preceded by a flattened head that it was holding up about four feet high. Based on its disturbance, the men estimated there might have been another forty feet of length obscured beneath the depths. In their separate statements, both M'Quhae and Drummond described the thing as being dark brown in color, but light brown to yellowish-white underneath. The captain discerned what looked like a "horse's mane" (or perhaps just seaweed) washed about the serpent's back, though the lieutenant had a different interpretation and felt the structure resembled a type of fin. The creature's snake-like body appeared to be about sixteen-inches in diameter.

By early October, when the *Daedalus* arrived back in England, word of the strange incident was published in the London newspapers and it created a swirling scientific debate. Based on the impeccable character and experience of the sailors involved, it was inconceivable they had been mistaken – or worse, had for some reason invented a story that would surely jeopardize their military careers. For a time after, sea monsters were all of the rage, as the press ran countless stories (which no doubt helped to stimulate newspaper sales), and a number of other experienced sailors came forward with their own, similar accounts that seemed to validate Captain M'Quhae and his crew.

One unabashed critic of the *Daedalus* account was none other than Sir Richard Owen, one of the preeminent naturalists of his time. Owen published a lengthy letter in *The Times* newspaper explaining how M'Quhae and the others had clearly misidentified a known animal in an unusual circumstance – a large, wayward seal, which had perhaps floated north on an iceberg that had since melted away. His argument was based on reconstructive illustrations of the sea serpent that had appeared in the papers, as well as his assumption that even experienced navigators might misinterpret a known but unexpected animal. Owen nominated two potential candidates – the leopard seal (*Hydrurga leptonyx*) and elephant seal (*Mirounga leonina*), the latter of which can grow up to twenty feet in length. He explained how the exaggerated size estimate of sixty feet might have been due to the wake the animal had created. Furthermore, Owen aptly pointed out the crew of the *Daedalus* had not observed any undulating motions by the animal – quite in contrast to other sea serpent reports. The famed zoologist concluded that sea serpents were merely a myth and could not exist, since no remains of them, fossilized or otherwise, had ever been found.

For his part, M'Quhae stuck to his proverbial guns and wrote a rebuttal to Owen, stating he and his crew had not made any errors in judgment, that they had all had ample observation time in order to form their opinions, and that they had merely described a creature that none of them had ever seen before. M'Quhae evidently had not been familiar with the writings of Erik Pontoppidan, so it is tantalizing he described the dark brown color, light underside, and horse-like mane so prevalent in the Norwegian legends. A few experts other than Owen published conflicting opinions on the matter, suggesting the so-called sea serpent was in all probability either a surviving, prehistoric, marine reptile known as a plesiosaur, a giant squid, rorqual whale, or perhaps even just a mass of floating seaweed.

With the episode unresolved a decade later, *The Times* published a correspondence from a third (as yet unidentified) officer of the *Daedalus,* who reminisced, "We all felt greatly astonished at what we saw, though there were sailors among us of thirty and forty years standing, who had travelled most seas and seen many marvels in their day. The captain was the first to exclaim: 'This must be the animal called the sea-serpent,' a conclusion we all at last settled down to."

Animal described by two zoologists from the deck of the HMS Valhalla
(Public Domain)

HMS VALHALLA SIGHTING

ONE COULD HARDLY HOPE FOR A BETTER OBSERVATION OF AN unknown animal than that which occurred at approximately 10:15 a.m. on the morning of December 07th, 1905. The spectators were both distinguished Fellows of the Zoological Society of London – specifically, respected ornithologists who were sailing about fourteen miles off the coast of Brazil on a scientific expedition. At the time, Edmund Meade-Waldo and Michael John Nicoll were guests on the yacht *HMS Valhalla*, which belonged to British nobleman the Earl of Crawford, an enthusiastic patron of the sciences.

As he leaned against the railing on the starboard poop deck, Nicoll suddenly exclaimed, "What's that?" and pointed at a great, rectangular, fin-like object that rose up out of the water, some fifty to one hundred yards distant. Meade-Waldo hurriedly raised his fine, German-made binoculars and also observed the structure, which apparently looked "soft… almost rubber-like," and of a color that was "dark, seaweed-brown." As best they could tell, the frilled fin was about four to six feet in length and extended from eighteen inches to two feet above the surface. It didn't resemble anything they had ever seen before. Suddenly and without warning, a small head, attached to a long neck, breached the surface, noticeably in front of the fin, by at least a foot and a half – and rose up between six to eight feet above the water, twisting back and forth. Both men later recalled the neck was about the thickness of a man's thigh and brown on top, though a dirty, white color underneath. Their initial impression was that the head looked turtle-like. Yet, Nicoll later wrote, "I feel sure, however, that it was not a reptile that we saw, but a mammal... the general appearance of the creature… gave one this impression." In a 1929 letter to sea serpent investigator Rupert T. Gould, Meade-Waldo opined, "The creature seen from the *HMS Daedalus*… might easily be the same."

Being that the craft had been traveling at least eleven knots when the creature was sighted, the *Valhalla* was unable to stop. But once Lord Crawford had been notified, the vessel turned about and made two subsequent passes near the same coordinates. At around 2:00 a.m. the following morning, three crew members heard loud splashing, and in the moonlight felt they saw a large animal moving just beneath the surface. In retrospect, it's fun to imagine what the late Sir Richard Owen might have thought of two, trained scientists describing a *Daedalus*-like animal swimming about. No doubt, he would have suggested it was merely an Amazon river dolphin *(Inia geoffrensis)* that had been washed out to sea!

MORGAWR
THE COUNTY OF CORNWALL IN FAR SOUTHWEST ENGLAND
extends into the Atlantic Ocean like a great finger. I mention this because if one were to look at a map of the North Atlantic, focusing on the Norwegian coast, Greenland and New England, where an abundance of sea serpent accounts have been logged throughout the centuries – one would be inclined to contemplate whether the adjacent coasts of western Ireland and England might produce similar reports. In fact, they do, with undeniably the most publicized monster being known by the Cornish name 'Morgawr' (the locals prefer the name 'Fessie'), said to periodically appear in the waters around Cornwall's Falmouth Bay. The stories date back to a dubious, 1876 newspaper article that recounts how some fishermen found a "baby serpent" coiled upon a buoy marking their lobster pots. The men were said to have captured the creature, but then after dragging it to the shore, "for exhibition"– tragically bludgeoned the poor thing to death, unceremoniously tossing its lifeless body back into the ocean.

Now it should come as no surprise that numerous newspapers have constructed sea serpent tales throughout the years in a shameless attempt to sell copies. The immense public interest in the *Daedalus* sighting did not go unnoticed by publishers looking to juice up circulation when their news cycle became slow. However, one bona fide exception may be the *Falmouth Packet*, a weekly Cornish tabloid that has in earnest attempted to raise awareness about their local monster, though admittedly, the *Packet* may have also been duped by hoaxers on more than one occasion.

According to letters to the editor, as well as articles featured in the *Packet*, the apex of Morgawr activity occurred during the mid 1970s, specifically the period between late 1975 and 1976. Although subsequent investigations revealed at least some of the reports and even a couple of alleged photos of the beast that were submitted to the *Packet* were contrived by a local prankster named Tony "Doc" Shiels – a self-proclaimed wizard, magician, artist, and entertainer who seemed to get his jollies by perpetuating the Morgawr myth. Still, some of the reports from that time may have merit.

One highly publicized incident involved locals George Vinnicombe and John Cock, who were fishing about twenty-five miles south of Lizard Point (a seemingly appropriate name) during the first week of July, 1976. The conditions at the time were apparently "Beautiful, calm and clear." At one juncture, George happened to glance off the starboard side of their vessel and noticed a large object resembling an upturned boat in the distance. Fearing people may be in danger, the two men approached cautiously. As they drew closer, the befuddled men realized the object was not manmade, was displaying at least two humps, and might in reality be the floating remains of a dead whale or other animal. Then quite suddenly, a seal-like head attached to a long neck rose straight up, three feet in front of the two bulges, looked straight at George and John, and then sank straight down and out of sight. At that moment, both men realized they had come across something they had never encountered in their four decades at sea. Both later characterized the creature as some sort of "prehistoric monster," at least eighteen to twenty-two feet in length, with a black body (grayish-looking under the water) and weighing several tons. Upon returning to port, Vinnicombe and Cock were reluctant to tell anyone what they had seen. But word of their startling encounter quickly got around Falmouth, and eventually the two men even appeared on some television shows, where they related the details of their encounter in convincing fashion.

A compelling Morgawr account from July 10th, 1985 involves local author and historian Sheila Bird. The animal in question was also observed by Sheila's brother, Eric, a scientist on holiday from Australia, as well as another couple. While sitting on a cliff near the village of Portscatho and enjoying a view of the "exceptionally calm" bay at around 8:00 p.m. in the evening, Eric leapt to his feet excitedly and began to point at a huge animal that was slowly swimming below them, about two hundred meters offshore. Definitively not a seal, nor any recognizable fauna, the thing was at least seventeen feet long in their estimation. As Sheila later described it in her letter to the *Falmouth Packet,* it was "An unfamiliar, marine creature with a long neck, small head and large hump protruding high out of the water, with a long, muscular tail visible just below the surface." Quite fortunately, a passing couple had a pair of binoculars on hand, so all of the witnesses were able to observe that the monster's flesh was "gray and mottled" and also displayed dark lines that resembled skin folds. To the surprise of all, Morgawr eventually sunk straight down like a rock and out of sight, identical to what local fishermen George Vinnicombe and John Cock had described in their report.

Morgawr sightings have sporadically continued throughout the years. On May 16th, 2000, holidaymakers Derek and Irene Brown were looking out at the sea near Falmouth when a weird object about two hundred yards offshore caught their attention. Before long, they could make out a series of humps, perhaps fifteen feet in length – and then a flexible, periscope-like object that broke the surface, which they took to be a head and neck. Irene dashed to grab a camera out of their vehicle, but by the time she returned, the thing had submerged. The couple waited around for an hour, but the entity never resurfaced.

As recently as 2019, Cornish residents were in a panic due to a mutilated dolphin that had been found washed up on the beach near the village of Harlyn. Based on enormous teeth marks that were visible on the carcass, some enormous creature had clearly bitten the animal in half. While a few marine biologists suggested that global warming may be causing great white sharks *(Carcharodon carcharias)* from southern waters to move into areas where they have never been documented before – one has to wonder if instead, monstrous Morgawr may have simply returned to his old haunt.

The author investigating sea serpent accounts at Zandvoort on the coast of Holland

CHESSIE

ON THE NORTH AMERICAN COAST, THERE HAVE BEEN surprising accounts of sea serpents in the modern era. Undoubtedly, the most venerable of these creatures is known as 'Chessie' – the monster of Chesapeake Bay, which is known to navigate that inlet between Maryland and Virginia. While the earliest Chessie anecdotes date back to the 1930s, events truly ignited during the summer of 1978, when a large, gray, snake-like animal was beheld by a number of eyewitnesses at various points throughout those waters. In one notable incident, a retired CIA agent named Donald Kyker and three neighbors from the community of Bay Quarter Shores at the mouth of the Potomac River watched four of the creatures swim by about seventy-five yards offshore. The critters were dark-colored and between twenty-five to thirty-six feet long, with the general appearance of "undulating, self-propelled logs." Kyker's neighbor, Howard Smoot, remarked the serpents were about eight inches in diameter and had heads shaped like footballs. He even shot at one of the things, causing them to submerge.

Videos of alleged sea serpents are unequivocally rare. But on May 31st, 1982, a resident of Kent Island named Bob Frew managed to capture footage of Chessie as the beast swam by his dock at Love Point. Several other witnesses, including his wife Karen, also saw the animal, which was described as resembling a twenty to forty foot telephone pole that undulated vertically just beneath the surface of the water. So compelling was the three-minute video that it prompted a scientific review by an assemblage of seven experts at the *Smithsonian Institution.* Director of Vertebrate Zoology and world-renowned herpetologist, George R. Zug, noted the film was too indistinct for him to form a conclusion. Yet, he did admit the footage appeared to show some sort of "animate" marine animal that he was not familiar with. Bob Frew ultimately became a local folk hero for his efforts, and in 1985 the State of Maryland even considered passing a state law in order to protect Chessie from harm. According to Dr. Eric Cheezum, a local historian with a lifelong interest in the Chesapeake Bay Monster legend (and one who's currently writing a book on that very subject), "Other than a very few dubious reports, there really haven't been many proper sightings since the early 1990s."

CALIFORNIA'S SEA SERPENTS
IT'S INTRIGUING THAT THE LION'S SHARE OF SEA SERPENT accounts stems from the Atlantic Ocean, particularly from its frigid northern waters. Still, we do find similar narratives arising from parts of the Pacific, in particular within certain bays on North America's western coast. Some anecdotes are exceptionally colorful – such as the legend of the 'Old Man of Monterey Bay,' or 'Bobo,' as he is affectionately called. In many ways tales about Bobo, which mostly date back to the 1920s and involve the crews of local sardine boats, are reminiscent of the classic sea serpents with one striking exception – Bobo is said to possess a face that resembles an ill-favored, old man, or in some renditions, a monkey! Nevertheless, some compelling, modern reports of Pacific Coast sea serpents do exist.

Halloween of 1983 will always be a memorable day for a nine-person road construction crew that was working at Stinson Beach, just north of San Francisco. According to a 2017 interview, which investigator Micah Hanks conducted with then crew leader Marlene Martin, it had been right around 2:00 p.m. in the afternoon when she received a radio call from one of her flagmen. The worker was alerting her, as well as fellow crewmember Matt Ratto, that there was something extraordinary in Bolinas Bay, about a quarter of a mile offshore – and that it was headed their way! Matt and Marlene peered out and could discern a massive wake headed toward the beach. As they watched in utter amazement, the culprit began to slowly turn one hundred yards offshore and head back out to sea. Both witnesses could see it was some kind of living creature – flat black in color and perhaps one hundred feet in length. Ratto managed to expeditiously procure a pair of binoculars, and the stunned onlookers were soon joined by the flagman as well as the rest of their curious road crew. Taking turns with the field glasses, the group was able to make out more details. Witness Ed Bjora later recalled the monster looked like a huge eel and may have been traveling as fast as forty miles per hour. Matt Ratto stated that at one point, he could clearly make out three distinct humps on the beast's back. Marlene Martin claimed she saw the creature open its mouth. She could even see its mouthful of sharp, crocodilian-like teeth. The great serpent thrashed around in the distance for a while before they lost sight of it. As you might expect, the number of observers involved in the Stinson Beach Incident spawned a great deal of interest from the press at the time.

Just over fifteen months later and a few miles to the south, twin brothers Bill and Bob Clark would experience their own altering experience. San Francisco Bay is perhaps one of the most famous harbors in North America and contains such famous landmarks as the Golden Gate Bridge and the island prison of Alcatraz. On the morning of February 02nd, 1985, the Clark Brothers were parked by the water enjoying the view, as well as some aromatic coffee. Some playful sea lions out in the bay caught their attention. Then suddenly, things took a frightful turn when those sea lions appeared to take flight from an unseen assailant. Without warning, what appeared to be the head of a long-necked creature rose up about ten feet out of the water and struck at the panicking pinnipeds. Bill and Bob continued to stare in awe as the action moved directly toward their position, since one of the sea lions was heading their way, with the mystery beast in hot pursuit! Finally, about ten yards in front of the Clark brothers, near a manmade, concrete harbor wall, the enormous serpent flexed and rolled just beneath the surface before heading back out to deeper water. Their overall impression was that of a titanic snake, at least sixty feet in length. When the sea serpent had momentarily paused, its body bunched up, and they could discern four humps rise up on its body.

Accused by some of fabricating a story, Bill and Bob Clark would become engrossed in their own personal journey of vindication and discovery. They've spent the better part of the subsequent decades trying to prove San Francisco Bay is occasionally frequented by an unknown creature. During 2004, the intrepid monster hunters even videotaped a line of distant objects out in the bay that they interpreted as being the humps of their sea serpent, however the resulting footage is controversial. Skeptics say the film merely shows a group of flocking water birds, though the dark objects appear to ultimately vanish below the surface, rather than fly off.

The Clark brothers first contacted me back in 2011 in order to share some of their extensive research, including some other eyewitness reports they'd collected through the years. Most recently in November 2014, Bill and Bob related to me how an ex-Marine pilot claimed he'd been driving across the Carquinez Bridge when he noticed a large, black hump in the bay below. The man was firmly convinced what he had seen was an unfamiliar creature.

The Naden Harbor Carcass, pulled from a whale's stomach in 1937 (Public Domain)

CADBOROSAURUS

MEANDERING UP THE PACIFIC COAST FROM SAN FRANCISCO Bay to Canada, one will eventually arrive at cozy Cadboro Bay, on the southern end of British Columbia's Vancouver Island. Here you will find a treasure trove of accounts relating to a celebrated local sea monster known by many colorful names, including 'Cadborosaurus,' 'Caddy' (for short), 'Amy,' and 'Penda.' The number of documented reports is striking, and while there are native legends and stories dating back to the nineteenth century, Caddy first made newspaper headlines a mere five months after her Scottish cousin, Nessie, during October of 1933. A sighting by a respected public official and his family that month got the ball rolling, and the editor of the *Victoria Daily Colonist,* a man named Archie Wills, enthusiastically promoted the local sensation as other witnesses began to come forward. The descriptions fit perfectly into the paradigm of sea serpent lore. Caddy was characterized as being some sixty to eighty feet in length, snake-like in form, and producing a series of large humps or "loops" when she undulated through the current. Perhaps the most riveting facet of the mystery is that observers through the years have consistently described Caddy's head as looking like that of a horse or camel, often adorned with horns, whiskers, or a mane of hair running down the neck.

In one especially dramatic incident, which occurred on January 18th, 1934, a young man named Cyril Andrews was duck hunting with his friend Norman Georgeson at a place called Gowlland Head. They'd decided to row a punt (flat canoe) just offshore in order to retrieve a wounded duck that was floundering in a kelp bed. As Cyril later put it, "About ten feet away, out of the sea arose two coils. They reached a height of at least six feet above me, gradually sinking under the water again, when a head appeared. The head was that of a horse, without ears or nostrils, but its eyes were in front of its head, which was flat just like a horse… To my horror it gulped the bird down its throat. It then looked at me, it's [sic] mouth wide open, and I could plainly see its teeth… which were like those of a fish… Shortly after, it sunk beneath the surface."

Sightings of Cadborosaurus have not been restricted to Cadboro Bay. In the decades since the 1930s, they have also been logged in the surrounding waters of the San Juan Islands and Strait of Georgia – and truthfully, all of the way from Alaska down to Washington State. During the 1970s, the phenomenon even attracted the attention of a Canadian scientist named Paul LeBlond, who specialized in oceanographic studies. LeBlond and some colleagues ultimately spent years documenting accounts of Caddy, as well as coauthoring two books on the subject. For the popular television series *Arthur C. Clarke's – Mysterious World,* LeBlond and marine biologist John R. Sibert traveled to the tiny village of Sechelt, British Columbia in order to interview a young man named John Andrews (no relation to Cyril). Andrews claimed he had observed an unknown forty to fifty-foot, "snakelike thing," swimming just below the surface, adjacent to a dock he was fishing from. He described the creature as being about as thick as his thigh, with two pairs of tiny fins and huge, cat-like eyes that could move in opposite directions like a chameleon's. When it swam, the thing undulated up and down.

Another notable incident involved floatplane pilots James Wells and Don Berends. On July 14th, 1993, the two aviators were practicing maneuvers in Saanich Inlet when they observed two, thirty-foot serpentine animals moving through the water at speeds approaching forty miles per hour. The animals created high loops with their backs when they swam. Both pilots were quite familiar with a wide range of local marine animals, including whales and seals, and stated this was unlike anything they had ever seen before. As recently as August of 2013, a man named Aiden Girard had a Caddy sighting at Cadboro Bay. According to Girard's testimony, the beast he encountered was gigantic, with a serpentine body that at times rose two feet out of the water. In his estimation, it possessed a head like a seal or horse and seemed to be looking around curiously.

As the staunch critic will quite correctly point out, virtually all of the evidence for the existence of sea monsters seems to be in the form of anecdotes and eyewitness testimony. And no matter how credible an eyewitness may sound, an account remains, in scientific terms, inconclusive. What science truly requires for formal description is physical evidence – a body (or at least a part of one). In point of fact, there seems to have been such proof within grasp on at least one occasion. Yet sadly, this priceless artifact has been lost to time.

Naden Harbor in British Columbia's Queen Charlotte Islands lies about six hundred miles northwest of Cadboro Bay. During the summer of 1937, there was a whaling station there where so-called 'flensers' would clean and gut the massive cetaceans that had been harpooned by local whalers. One day as these hardworking laborers (mostly Asian immigrants) were cleaning out the stomach of a sperm whale, an incomprehensible creature slid out. The monstrosity was about twelve and a half feet long, snake-like, and had a head similar to a camel's, with two flippers and a fluke tail. The carcass looked wholly intact, and no one who was present had ever encountered anything like it, despite many years of disemboweling whales. The man in charge , F. S. Huband, was alerted at once, and when he laid eyes on the thing, he ordered it immediately be photographed. The serpentine atrocity was then outstretched on a long table. Wooden boxes laid on end were used to extend the setup. A white sheet was laid underneath and just behind the animal, in order to provide contrast. Although tragically, its remains were discarded soon after, so all that remains are the photos, which have since been published in newspapers and books.

Provided the images are genuine, the Naden Harbor carcass does not match any known species, even from the fossil record. Yet it does correspond admirably with the countless descriptions of sea serpents from around the world. Albeit, based on the relatively small size of the subject, it would be fair to assume the specimen was undersized. While the suggestion has been made that the carcass merely represents a baleen whale fetus, decomposed basking shark, or mangled elephant seal, a few scientists have expressed the view that the photos deserve a closer look. Typically skeptical paleontologist Darren Naish acknowledges there's a chance the Naden Harbor carcass might represent "a highly unusual, apparently new type of unusual vertebrate animal." Although Naish has also stated the possibility of a misidentification of a decomposed basking shark carcass cannot be totally ruled out.

❦ 4 ❦

Ogopogo & Other Canadian Lake Monsters

"His mother was a polly while his father was a whale." -"OGO-POGO" – PAUL WHITEMAN ORCHESTRA – 1924

In the last chapter, we examined a celebrated sea serpent with the whimsical name Cadborosaurus, whose existence is seemingly supported by an actual physical specimen (carcass) documented near the coast of British Columbia, Canada, in 1937. Traveling inland from Cadboro Bay just one hundred seventy miles, one arrives at picturesque Lake Okanagan. Narrow in design, measuring some eighty-four miles long by three miles wide, Okanagan shares affinities with Scotland's Loch Ness. It's a cold-water, glacial lake that has almost the exact same maximum depth as Loch Ness at 761 feet and is inhabited by several of the same fish species, including trout, salmon, and northern pike. Both lakes were formed in the same way, at around the same time about ten thousand years ago – and both lie in northern latitudes that are within eight degrees of each other. Like Loch Ness, Lake Okanagan is also joined to the sea, in this case by the mighty Columbia River via a meandering tributary called the Okonagan River. And similar to Loch Ness, Lake Okanagan is said to be home to a famous monster.

So memorable in fact, it has inspired both a song, as well as a commemorative Canadian postage stamp! Known popularly as 'Ogopogo' or 'Ogie,' for short, its routinely mentioned physical similarities to Nessie and our other so-called sea serpents around the globe cannot be ignored.

NAITAKA

AROUND THE LAKE OKANAGAN AREA THERE IS A ROBUST NATIVE tradition that dates back centuries and tells of a terrifying lake demon that perpetually has to be appeased with sacrificial animals. The indigenous Okanagan tribe refers to the beast as Naitaka (pronounced N'ha-A-itk) and it is firmly believed to inhabit an underwater cave between Rattlesnake Island and Squally Point. According to an oft-repeated legend, when Indian braves would canoe past Squally Point, they would frequently observe the rocky shore was blood-stained and littered with the ravaged bones of Naitaka's victims, be they animals or men. Only by depositing a live, sacrificial pig, dog, or chicken into the depths was one allowed to navigate the lake safely, without fear of reprisal from the monster.

One old story from 1854 features a local named John McDougall who was traveling the lake in a canoe with his two, swimming horses in tow. About halfway across, McDougall realized he had forgotten to bring along the standard offering of a small, sacrificial animal, but it was too late to turn back. Suddenly, a tremendous unseen force seized his pair of stallions and dragged them below the surface. Thinking his canoe might be capsized, McDougall quickly cut the rope. The horses evidently never resurfaced. It soon became obvious to the shaken man that the slighted Naitaka had collected her toll.

An intriguing piece of evidence that seems to fortify Ogopogo's reality is there have apparently been ancient pictographs found around the lake, which seem to depict an unusual animal with a horse-like head, long body, and bilobate (double-lobed) tail. The primordial rock art doesn't match any known creature, though curiously does present a curious facsimile of our Scottish Water Kelpie – and in its tail section – a whale.

A MIsPLACED MANATEE
ACCORDING TO CRYPTOZOOLOGIST AND AUTHOR JOHN KIRK who has investigated the Ogopogo phenomenon for years, one of the frustrating aspects of the mystery is the fact that through the years, pranksters have planted whalebones, small shark carcasses and other items around the lake in an attempt to fool the public. In one utterly weird case from 1914, a huge, dead animal cadaver was found on the shore. Characterized as weighing some four hundred pounds, having a round head and flat tail, a local naturalist determined the specimen was most likely a Florida manatee *(Trichechus manatus)*. The problem is that manatees, also known as "sea cows," are tropical, coastal species that live well over two thousand miles away from Lake Okanagan. It is unclear how such a creature ended up at the lake, unless it had been put there deliberately.

1920s – OG0P0G0 Is B0RN
THERE'S LITTLE DOUBT THE LAKE DEMON HAS BEEN A FAMILIAR trope around the Okanagan area as far back as anyone can remember. However, there was something about the roaring 1920s that thrust the monster into the spotlight, primarily driven, it would seem, by a marked uptick in sightings. It's worth noting this significant flap predated events at Loch Ness by over a decade. And as you might expect, local boat captains were the first to notice the thing. Throughout 1923, Matt Reid – skipper of the steamship *S.S. Okanagan,* claimed he would occasionally spot the creature disporting itself near Squally Point. He described the animal as being about forty feet long, with a head like a horse. Captains from

other vessels were known to blow their horns when they caught sight of the beast.

Then, in 1924, two British composers wrote a catchy, dance hall song, which was adapted two years later by an Okanagan area musician, who tweaked the lyrics so that the tune paid tribute to the monster. "Ogopogo" quickly supplanted the hard to pronounce First Nation moniker. Local newspapers soon got involved in the act, eagerly printing stories about the creature whenever a report surfaced. When the national newswires circulated some of the reports, self-proclaimed monster hunters from all over the place arrived on the scene, hoping to capture the ultimate prize – which they ultimately failed to do. Still, the year 1926 was a particularly active period for encounters. At 7:00 a.m. on the 19th of July, motorist John Logie claimed he raced alongside the beast as it propelled itself through the water adjacent to the shoreline at speeds approaching twenty miles per hour. Yet, the seminal sighting occurred on November 18th, when as many as fifty to sixty people attending a baptism on the shore watched Ogopogo swimming in Gellatly Bay, a mere fifty yards away. One parishioner, a Mrs. Clifton, described the unexpected animal as displaying a sheep-like head followed by two, tire-sized loops.

SNAKY SIGHTINGs CONTINUE

SIMILAR TO THE SITUATION AT LOCH NESS, ACCOUNTS OF Ogopogo have steadily continued throughout the years, though there seem to be decades like the 1920s that produce more sightings for whatever reason. In a 1938 letter addressed to Sir Edward Mountain, who'd organized the very first Loch Ness expedition four years earlier, a vacationing Englishman named Leslie Casar described seeing the monster of Okanagan. Casar wrote the creature had been some twenty to thirty feet long, with a humped body, a long neck, and a horse-like head. The thing evidently had displayed impressive speed. On July 02nd, 1949, eight people spotted Ogopogo around 7:00 p.m., just as the group of them were starting to take a boat out onto the lake. The witnesses included Mr. & Mrs. L. L. Kerry, Mr. & Mrs. W. F. Watson, their children Bobby and Janey, as well as Dr. and Mrs. Underhill. After peering through binoculars for several minutes, Dr. Underhill concluded there appeared to be two of the animals swimming together. Both beasts were around thirty feet long, with undulating, coiled bodies – and they occasionally displayed their forked tails. An incident on June 29th, 1950 involved no fewer than seventeen observers, including the driver of a Greyhound bus, Gordon Radcliffe, and all sixteen of his passengers. About three miles north of the town of Summerland, Radcliffe noticed a disturbance in the lake and pulled his bus over to the side of the road in order to investigate. At least twelve feet of the mystery creature were showing above the water. It appeared to be brownish in color and possessed a flat head that it moved from side to side. On the morning of May 19th, 1951, two businessmen fishing at a place called Bear Creek watched an odd commotion in the water for some ten minutes, until a horse-like head appeared briefly and then submerged. The men attempted to chase the varmint's visible wake in their boat, but couldn't keep up.

Statue of Ogopogo at Kelowna, British Columbia (Wikimedia Commons)

THE 1960s – OGOPOGO AT THE GO-GO
THE DECADE OF THE 1960S WAS A BUSY TIME FOR THE "SNAKE in the Lake," as it was sometimes called by the locals. On Sunday, July 29th, 1962, the monster was seen by an entire group of church picnickers from the shore at Kelowna City Park. Parishioner Ernie Cowan recalled the dark creature, which was at least a dozen feet long, undulated through the water about three hundred yards away. Mrs. Marie Vetter confirmed the thing looked serpentine and its head had been "horsey-looking." In July of 1967, school teacher Brenda Briese was piloting a speedboat that was towing a water skier along the shore, when she noticed hordes of tiny fish were flopping frantically out of the water ahead of her. Looking slightly to her right about one hundred feet distant, Brenda noticed something which she at first took to be a floating log: "I saw twenty feet of it, a curved

shape with no head and no fins, just like a whale, resting under the surface of the water." The startled woman actually had to swerve her boat in order to avoid hitting the object, which sank quickly down and out of sight. That same month, Ogopogo was spotted by fifteen cannery workers in the community of Naramata, as well as by Mr. & Mrs. John Durant, who claimed they observed the animal as it spouted like a whale. Taxi driver Paul Pugliese had a well-publicized encounter on September 07th, 1968. As he later recalled on a couple of network TV shows, including *In Search of* and *Arthur C. Clarke's Mysterious World*, Pugliese stated he was driving down Abbott Street when he happened to glance out at the lake. He was surprised to see a horse-like head pop up out of the water. Pugliese pulled his cab over to the side of the road and jumped out in order to get a better look. All in all, he felt it looked like a "big serpent." After a few moments, it submerged, creating big waves.

THE FOLDEN FILM

BECAUSE THE 1960S SEEMS TO BE ONE OF THE MOST ACTIVE periods for Ogopogo sightings, it's understandable that it's also when the first, compelling motion picture of the beast was filmed. The highly analyzed and discussed footage was captured on a warm August day in 1968. Sawmill worker Art Folden of Chase, British Columbia was with his family, returning from a visit to the city of Penticton at the southern tip of Lake Okanagan. As they motored north along Highway 97, Art looked out across the calm lake toward Squally Point and spotted something immense churning through the water. Folden evidently exclaimed to his wife, "There's Ogopogo!" to which she laughed. Nevertheless, Art pulled their car over to the side of the road and grabbed an 8mm movie camera that he always carried along with him. Fortunately, he also had a telephoto lens attached to his camera, since he was on the edge of a high bluff and the creature was three hundred yards offshore. Art began to film and ultimately captured about eight hundred frames (a little less than a minute of footage) of a massive, dark, hump breaching the surface three different times and then submerging, while it lunges ahead. As he was low on film at the time, Art only shot when the hump was visible and took his finger off the camera trigger whenever the object submerged. An impressive amount of swirling white foam can be seen at the front of the creature, which Art later estimated to be sixty feet long (investigators on a 2005 TV show called *Is It Real?* determined the subject in the film was probably only about half that length). When Ogopogo submerged for the final time, Art and his family waited around for fifteen to twenty minutes in order to see if the monster would reappear. When it didn't, they jumped in the car and headed home.

Like Nessie photographers Hugh Gray and Peter MacNab before him, Art Folden apparently was reticent to go public with his film for some time after the incident. Although he had grown up near Lake Okanagan and had always felt there probably was some truth behind the Ogopogo legend, Art was also aware his film might be viewed by many as some sort of joke. Word got out however, and in 1970 a copy of the film was obtained by Art's stepbrother and screened before influential members of the Kelowna City Council. Reactions were mixed. At least one member in attendance felt the film was inconclusive. But other viewers, including lifelong residents of the area who had grown up with the Ogopogo legend, were convinced the stories were now validated; clearly there appeared to be a truly, gigantic animal in the lake – something which should not be there.

TOUCHING A DRAGON

FILMING A SEA MONSTER IS ONE THING; HAVING ACTUAL physical contact with one is another matter, altogether! Yet, that is what apparently occurred to a young woman named Barbara Clark in July of 1974. Barbara had grown up in a small community on the southern edge of Lake Okanagan. That particular summer morning, she'd decided to swim out to a floating platform about four hundred yards offshore. Just as she was approaching the attached pontoon, Barbara felt something massive and unseen swim by and scrape her leg. The startled girl lunged onto the platform, and looking down at the water saw an eight-foot, "dull, dark gray" hump extending out of the water. Just beneath the surface, trailing about ten feet behind, she also observed a tail that "was forked and horizontal like a whale's and... 4' to 6' wide." The immense creature slowly swam in an undulating fashion away from Barbara, much to her relief. When she saw it surface again in the distance, Barbara dove off the platform and swam like Olympian Katie Ledecky back to the shore. In a 1986 letter to cryptozoologist Richard Greenwell, Clark recalled, "The thing looked more like a whale than a fish, but I have never seen a whale that skinny and snaky-looking before."

THE FLETCHER PHOTOs

VANCOUVER BUSINESSMAN ED FLETCHER DECIDED TO SPEND the summer of 1976 vacationing at Lake Okanagan. Ed had acquired a souped-up speedboat, perfect for water skiing and just plain going fast. On August 03rd, as he and his daughter Jill were idling, "About two blocks out" in Gellatly Bay, they were shocked to see an immense, serpentine form rise up directly in front of them – writhing and twisting. In Ed's estimation, the smooth, brown, back of the massive creature was close to eighty feet in length. He even managed to snap a photo, which superficially resembled rolling waves, though a closer look at the top edge of one bulge revealed a serrated edge. As it turned out, that particular afternoon ended up being quite eventful, since Fletcher, along with both of his daughters and a couple of family friends, ultimately chased Ogopogo around the lake for an hour or so. The group claimed to have watched the monster surface twelve to fifteen more times, while managing to snap four additional photos. Ed later stated the animal had "skin very similar to a whale's." In time, Ed's photo received a quite a bit of publicity, even getting him onto a TV show, where he explained how he felt Ogopogo was attracted to his boat because its motor produced a "high-amperage." Fletcher returned to Okanagan the following summer for further investigation, alleging at least twenty-five more sightings of the animal – and even insisting his boat was lifted up out of the water by the creature at one point! Later, Ed unsuccessfully attempted to claim a million dollar reward (insured by Lloyds of London) that had been offered for definitive proof of Ogopogo's existence.

THE THAL FILM

ON AUGUST 11TH, 1980, A VANCOUVER RESIDENT NAMED LARRY Thal, as well as his family, were visiting friends – relaxing on the beach at Bluebird Bay Resort. There were around thirty other people also relishing the gorgeous weather that day. After a while, some folks began to notice a mysterious wave at least one hundred yards offshore. At first, virtually everyone took it to merely be the lingering wake of a boat that had long passed. But the lake was otherwise calm and the disturbance remained for a while – moving back and forth and reversing directions at least three different times. Excitement ramped up as the stunned observers realized the cause seemed to be some immense, elongate creature undulating beneath the wave. Thal ran and got his 8mm movie camera and subsequently shot about ten seconds of the anomaly. Local Ogopogo investigator and author Arlene Gaal was on the scene within hours and obtained a sworn affidavit from at least sixteen individuals who had been present during the event. Gaal also acquired the rights to the film from Larry Thal. In Arlene's estimation, the animal involved had been at least sixty feet in length and had displayed some sort of appendages that broke the surface periodically.

CHAPLIN VIDEO

OGOPOGO WAS FEATURED IN ICONIC *TIME MAGAZINE* IN 1989.
On July 18th of that year, father and son Clem and Ken Chaplin had gone down to the water's edge, since Clem thought he might have sighted the monster at a particular spot the day before. Ken brought along a video camera. Sure enough, when they peered out across the lake, the men observed what they later described as a fifteen-foot, smooth-green, spotted, reptilian animal slithering along only sixty feet offshore. Ken managed to film several minutes of the creature, including a segment where it appears to slap its flipper on the surface before submerging and disappearing. The resulting footage and images made headline news and were widely published, even being featured on an episode of the hit TV show – *Unsolved Mysteries*. However, scientists who analyzed the film were immediately struck by the similarities between the animal Chaplin had videotaped and a common beaver *(Castor canadensis)*. The presumed flipper slap was in fact a dead ringer for the tail slap made by a beaver before it dives. Watching the clip in slow motion; one can even see two little beaver feet rise up before the tail slap. Needless to say, most Ogopogo investigators agreed with the beaver conclusion, though Chaplin continued to insist that there was no chance he could have been mistaken about what he had filmed.

DEMARA VIDEO
THE MOST DISCUSSED OGOPOGO FOOTAGE FROM THE 1990S was videotaped by a young man named Paul DeMara on July 24th, 1992. Interestingly, Paul's father, Monty, claimed to have had his own sighting of the famous monster near Squally Point back in 1951. At the time of his own sighting, Paul was sitting with some friends on a deck at his mother's home, overlooking the lake. All at once, Paul's mom noticed something unusual swimming by and yelled for Paul to grab his camcorder. As he began to shoot, DeMara could make out what appeared to be either one large or perhaps several smaller animals swimming just beneath the surface. Just

then, a speedboat towing a water skier came into frame. In the video, the boat actually appears to turn suddenly in order to avoid hitting the strange disturbance, causing the water skier to fall dangerously close to the action. The early 1990s were also notable due to the fact that a well-financed Japanese television crew mounted an elaborate expedition to search for Ogopogo. With help from local investigator Arlene Gaal, the TV crew deployed divers, a helicopter, and even a submarine – and felt they might have even captured their very own footage of the beast at one point.

CHAsED BY A SEA SERPENT

FOLLOWING A LONG HIATUS, OGOPOGO MADE A DRAMATIC reappearance in front of two astonished women on April 21st, 2015. Suzie St-Cyr Crowley and Marie Letourneau were cruising Lake Okanagan in Suzie's pontoon boat when Marie heard four, loud, abrupt sounds (like air escaping) directly behind them. Looking back, both women were stunned to see an interminable, dark, snake-like animal following behind their vessel. The following day, Suzie told a reporter for Kelowna's *Daily Courier* newspaper: "Usually a snake goes side to side. This one went up and down. You could see four or five peaks of its back... It was so long. Unbelievable." Letourneau, who was visiting from out of town at the time, had apparently never even heard of the Ogopogo legend. By the time she had managed to get over her initial shock (crying "Oh My God!") and grabbing her phone in order to try and take a photo, the creature had lowered its head and plunged beneath the surface, not to be seen again.

HALBAUER VIDEO

BROTHERS DAVID AND KEITH HALBAUER WERE STANDING ON A beach at West Kelowna on September 12th, 2018. It was a bright, sunny day. Perhaps too sunny, in fact, because when something surfaced around three hundred yards offshore, extreme glare prevented them from determining just exactly what they were looking at in the distance. Still, David grabbed his phone and attempted to film the object, which the men estimated to be some forty-five feet in length. "It was like a dinosaur, I guess. Like a giant, giant snake," David told a reporter the next day. The cylindrical, jet-black creature had surfaced and then rolled in the water for a few moments before submerging. Keith looked out just in time to see a number of large waves formed by the displacement of water. Despite the distance involved and brevity/shakiness of the video, the resulting footage was picked up by various news outlets and subsequently went viral on the Internet.

NEUDORF VIDEO

YET ANOTHER REMARKABLE VIDEO POTENTIALLY FEATURING Ogopogo was uploaded to the Internet as recently as January 02nd, 2020, although the footage was evidently shot a year and a half earlier on July 10th, 2018. Captured by sixteen-year-old Blake Neudorf on his cellphone, in my own opinion the footage presents extremely compelling evidence. At the time, Blake and his father were fishing from a dock in Kelowna. The day in question had been cloudy and overcast, but the lake surface remained flat and calm, indicating no wind to speak of. It is worth noting the sound of a helicopter can be heard throughout the duration of the minute and a half video, leading at least one skeptic to speculate the chopper was merely causing a phantom wave to form on the lake. But Blake later stated the helicopter had been behind him and his father, hovering overland the entire time. Be that as it may, young Blake began filming when the animal first appeared and continued throughout the duration of it crawling along the surface of the water, until it disappeared behind an outcrop. A seemingly enormous, long, dark 'whatsit' can be seen swimming along, with one to two or three humps rising up at various points – moving with the classic vertical undulations that have been described by numerous lake monster and sea serpent eyewitnesses. Neudorf estimated the creature was at least sixty feet in length and was a few hundred yards offshore. The audio captured on the video is also quite telling, as you can hear Blake repeatedly exclaiming, "What the heck is that?" and his bewildered dad answering, "I have no clue!"

MANIPOGO

BESIDES OKANAGAN, THERE ARE A MULTITUDE OF OTHER Canadian lakes that boast monster legends. Lake Manitoba in Central Canada stretches some one hundred twenty miles long by twenty-eight miles wide, though the water is uniformly quite shallow. Because Canadians are famous for their hearty sense of humor, Manitoba's mystery beast has been anointed 'Manipogo,' a play on Ogopogo, obviously. One of the oldest accounts dates back to the year 1909 and involves a trapper from the Hudson Bay Company named Valentine McKay who claimed to have seen a huge, lumbering creature swimming around near a place called Graves Point. Then in 1955 Joseph Parker, along with Albert Gott and his two sons, watched a gigantic hump rise out of the water in approximately the same vicinity. Two years later, a pair of men working on the shore described seeing a large, serpentine animal writhing about. During one notable incident that occurred on July 24th, 1960, twenty some picnickers at a popular park observed a multi-humped creature that displayed a flat head.

Other reports were to follow, prompting zoology professor Dr. James A. McLeod from the University of Manitoba to organize an expedition at adjacent Lake Winnipegosis. In addition to the reports, the scientist's interest had been bolstered by a claim from a man named Oscar Fredrickson, who stated he had discovered a non-fossilized vertebrae (backbone) six inches long in the lakebed of Winnipegosis during the 1930s. The mysterious artifact had supposedly been lost in a cabin fire some years earlier. But not before Fredrickson had managed to carve a wooden replica of the object, which bore a remarkable resemblance to the backbone of a whale. Two remarkable aspects being the closest ocean to Lake Winnipegosis lies at least four hundred miles away, and also Dr. McLeod felt the bone, if genuine, looked like it belonged to a prehistoric species. But investigator John Kirk has pointed out the region had long been inhabited by Icelandic settlers who might have brought some whale bones along with them. Regardless, McLeod evidently failed to find any more bones or any proof that Manipogo existed. There is, however, a curious photo taken by fishermen

Richard Vincent and John Konefell on August 12th, 1962, which seems to show a sizable, black, snake-like form with its midbody arching into a pronounced loop, swimming on the surface of Lake Manitoba. The two men claimed around twelve feet of the beast were visible from their boat, which was at least fifty yards away at the time. Ultimately, whatever the creature was, reports of Manipogo seem to have ceased following the 1960s, although there is now a popular lakeside park named after the monster.

The author investigating reports of Igopogo at Kempenfelt Bay, Lake Simcoe, Ontario

IGOPOGO

THE EASTERNMOST MEMBER OF CANADA'S 'POGO' CLAN IS alleged to dwell in Lake Simcoe, in the Province of Ontario. Nineteen miles long by sixteen miles wide, the body of water is relatively shallow by lake monster standards – with a maximum depth of only one hundred thirty-five feet. Also, unlike Okanagan and Manitoba, Lake Simcoe is nestled in one of Canada's most densely populated regions. Sightings of 'Igopogo,' also known as 'Beaverton Bessie' and 'Kempenfelt Kelly,' are relatively rare. According to one account published in *The Oakville Journal Record* – a minister, a funeral director, and both of their families were boating on the lake on July 27th, 1963, when their vessel was approached by a colossal, black, serpentine monster between thirty to seventy feet long. One witness described the creature as being about as thick as a stovepipe, with a head like a dog or seal. Then, in August of 1979 three women driving along the lakeside claimed to have seen an enormous, brown hump break the surface about a hundred yards offshore, head in their direction, and then submerge, after which they could make out a noticeable wake shoot off in the opposite direction. In 1983, there was reportedly a sonar reading of an unidentified "large animal" in the lake, and in 1991 a resident videotaped something in the water that one investigator described as resembling "a massive seal." I've personally collected one other vague Igopogo account. Ontario resident Leslie Carter related to me how her best friend's daughter had once spotted a "huge, snakelike thing" in the waters around Georgiana Island.

If one were to head due east from Lake Simcoe, in a little less than four hundred miles they would arrive at a vast body of water that is said to be the home to the most celebrated of all American lake monsters. It would in fact be fair to say that the legendary creature is unquestionably the champ!

5

The Lake Champlain Monster & Neighbors

"You can call it whatever you want. I'm telling you, in that lake there is something extraordinary." - SANDRA MANSI – 1992

Of all of the freshwater bodies that are said to conceal a famed monster, Lake Champlain is easily the largest. Sandwiched between the northernmost parts of New York and Vermont, with a tiny section that extends across the border and into the Canadian province of Quebec, Champlain is over one hundred miles long and eleven miles wide, with five hundred miles of coastline and a maximum depth of four hundred feet. Similar to Loch Ness, as well as Canada's Lake Okanagan, Champlain lies nestled in a valley carved out by glaciers during the Pleistocene epoch. The region was temporarily an arm of the Atlantic Ocean when those glaciers melted around ten thousand years ago. Unlike its counterparts, Champlain's waters do not always maintain a constant temperature, sometimes freezing over in winter and warming up significantly during the summer months. It's an extremely biodiverse lake, with a wide array of native species, though sadly has endured various ecological issues through the years, including pollution and related algae blooms, as well as some fifty invasive species that threaten the natural balance. In addition, the region boasts a rich history. Several significant naval battles were fought there in the nation's early years. Still, a substantial slice of the local culture revolves around the legend of the Lake Champlain Monster, popularly known as – 'Elsie,' 'Sammy,' 'Champy,' or most frequently, 'Champ.'

NATIVE AMERICAN HORNED SERPENTS
INDIGENOUS LORE HINTS TO CHAMP'S EXISTENCE. THE NATIVE Abenaki tribe of Lake Champlain's eastern region share a legend about a monstrous, man-eating, water serpent that is referred to as Tatoskok, which is said to sport a pair of horns on its head. Similarly, the Mohawk people of the western coast tout a mythical beast named Onyare, that is also portrayed as a horned, dragon-like, serpent – fond of capsizing canoes and devouring unfortunate passengers. Reminiscent of the aboriginal tales that surround Canada's lake monster Ogopogo, the Mohawk believed that offering a sacrificial animal to Onyare would avert certain tragedy. The description of horns is most interesting, as it draws parallels to the Canadian sea serpent Cadborosaurus, as well as some lake monster traditions from Europe. However, it's safe to assume at least part of this notion may have been inspired by rare sightings of swimming deer and moose.

DE CHAMPLAIN'S DEBACLE
ANALOGOUS TO THE LOCH NESS MONSTER LEGEND AND ITS aged account of St. Columba's dubious Nessie encounter, the Lake Champlain Monster is rarely mentioned without a reference to the lake's discoverer and namesake, French explorer Samuel de Champlain. It has often been stated that in the year 1609 while visiting the region, de Champlain observed a twenty-foot serpentine creature swimming in the lake. This dramatized incident is often touted as the first sighting of Champ by a white man. However, like

the St. Columba story, this narrative has been taken totally out of context – and the misunderstanding can apparently be traced back to an embellished magazine article from a 1970 periodical called *Vermont Life*. In truth, within de Champlain's journal from the time, he had simply written about how he'd observed the local natives catching a ferocious-looking fish called the Chaoufarou that was said to grow up to eight feet long and which was completely unfamiliar to him. Based on de Champlain's Chaoufarou description, which includes a double row of razor sharp teeth, it seems fairly obvious the animal in question was merely a frightening-looking species known as a longnose gar *(Lepisosteus osseus)*, although its size may have been exaggerated a bit.

CAPTAIN CRUM'S SEA SERPENT

THE FIRST WIDELY DISSEMINATED (THOUGH HIGHLY SUSPICIOUS) Champ account was published in a local newspaper called the *Plattsburgh Press-Republican* on July 24th, 1819. Our first clue the sensational story might be fabricated lies in the name of the primary witness, a "Captain Crum." According to the article, Crum was sailing a small dinghy about two hundred yards offshore in Bulwagga Bay when he spotted the black, undulating beast, which he portrayed as some 187 feet in length with a flat, fifteen-foot head, which the creature reared above the surface of the water. When we consider other curious details – that its back was "as large as a common potash barrel," that it had "eyes the color of a peeled onion," a red stripe around its neck, only three teeth and a "star in the forehead," it's fairly easy to write this one off as the product of an editor's imagination, probably inspired by the New England Sea Serpent flap two years prior. Still and all, we can't ignore the possibility that local knowledge of a strange animal in Lake Champlain may have played a contributing factor in the story.

CHAMP: WANTED DEAD OR ALIVE!

CHAMP'S TRUE COMING OUT PARTY HAPPENED DURING THE year 1873 when several notable sightings made big news. First, *The New York Times* reported how a railroad crew had encountered an enormous serpent with glistening, silver scales cavorting about in the lake. Then, in July there was an account that involved Clinton County Sheriff Nathan H. Mooney, who described seeing from a distance of fifty yards an "enormous snake or water serpent," which he estimated to be twenty-five to thirty feet long. In the most dramatic incident of that year, the steamship *W.B. Eddy* was cruising along when it apparently struck an unseen object, which caused the vessel to list. A few moments later, flabbergasted passengers reputedly watched as a horse-like head and neck rose up out of the water mere yards from the boat! Not to miss an opportunity to dazzle, the infamous showman P.T. Barnum, who'd caught wind of the dramatic events, quickly offered up a reward in the sum of fifty thousand dollars for the "hide" of the Champlain Monster, which he intended to exhibit at his Traveling World's Fair. Needless to say, Barnum's reward was never collected.

S.S. TICONDEROGA SIGHTING

APART FROM A 1915 *NEW YORK TIMES* ARTICLE THAT describes a multiple witness sighting, accounts of the Lake Champlain Monster immediately following the nineteenth century were infrequent. But during a bridge dedication in 1945, several passengers standing on the deck of the paddle steamer *S.S. Ticonderoga* claimed they observed Champ's head and neck momentarily rise up out of the surf. Two years later the animal was encountered by three fishermen, according to an article that appeared in the *Burlington Free Press* on September 20th, 1947. One of the anglers explained to a reporter, "Out of the depths reared a huge, dark form which moved swiftly in a northwesterly direction. Three segments appeared clearly discernible above the water's surface, separated one from the other by about five feet of water, the overall length of the creature being an estimated twenty-five feet… It moved with incredible swiftness – about fifteen miles per hour – and disappeared altogether in about two minutes." Ultimately, it would be almost three decades before Champ would make another big splash. But, it would turn out to be a doozy!

MANSI PHOTO
WITHIN THE FIELD OF CRYPTOZOOLOGY, THE SO-CALLED 'MANSI Photo,' which alleges to show the head, neck, and humped back of the Lake Champlain Monster, is widely considered to be the best photograph evidence of any aquatic cryptid ever captured. Yet, the picture is still viewed as being controversial after four decades.

On the afternoon of July 05th, 1977, Sandra Mansi, her fiancé Tony, and her two children from a previous marriage were traveling by car along the northeastern section of the lake, somewhere near the town of St. Albans, Vermont. The weather was evidently pleasant that day and the family had decided to drive a gravel road close to the lakeshore in order to enjoy the view. After parking at the edge of a field, the family continued down to the beach so the kids could wade in the shallows, while the adults sat and watched. At some point, Tony decided to run back to the car in order to retrieve his sunglasses, as well as a Kodak Instamatic camera. It was while he was away that Sandi noticed a churning on the surface of the water, perhaps one hundred fifty feet offshore.

Her first impression was that there was a huge fish or school of fish causing the disturbance. Then suddenly, an object broke the surface of the water and began to rise up. Sandra quickly determined it appeared to be the small head and long neck of some huge, prehistoric-looking animal. She later recalled that initially, its mouth seemed to be ajar. Eventually, the bulge of what appeared to be its back appeared behind the head and neck, which Mansi estimated to be six feet above the water. Her impression of total, visible length of the smooth-skinned creature was at least eight feet and possibly up to twelve feet. Sandra had grown up on Lake Champlain and remembered fondly how her grandfather would often tell cautionary tales about its monstrous resident. She knew at once she could only be looking at the legendary Champ. Within moments, Tony returned from the car, and according to all accounts, noticed the monster and began yelling at the children to get out of the water and hurry back up to the car. Fortunately, Tony also had the foresight to hand the camera over to Sandra, who snapped a single image of the creature before turning and joining in the panicked retreat. The entire encounter lasted no longer than four minutes.

As the family drove hurriedly away from the scene, they discussed whether they should report their remarkable encounter to the authorities, but ultimately decided they didn't want to subject themselves to any potential ridicule. Yet when the film from their camera was developed, they were amazed to behold the resulting image, which seemed to quite obviously show the profile of some type of antediluvian beast. Still, the family remained reticent about the incident, so the picture was tucked away in a family photo album for three years until Sandra took it to her workplace in order to show it to a colleague who had expressed an interest in the Loch Ness Monster. Word of the fantastic photo quickly found its way to Joe Zarzynski, a dedicated Champ investigator who didn't waste any time in contacting Sandra. When Zarzynski viewed the picture, he was amazed and sent copies to various academics for analysis, including zoologist George Zug at the Smithsonian Institution (he of the 1982 Frew/Chessie film analysis), as well as cryptozoologists Dr. Roy P. Mackal and Richard Greenwell. Greenwell, in fact, was on the staff of the University of Arizona at the time and managed to get the image both digitized and analyzed by Dr. B. Roy Freiden from the university's Optical Sciences Center. In a report dated April 30th, 1981, Freiden concluded the subject in the photo appears to naturally belong in its surroundings. That is, the picture was evidently not tampered with or manipulated in any way. Though, he couldn't conclusively rule out the possibility that someone had intentionally placed some type of fake monster out into the lake. One issue was the fact that Sandra Mansi had apparently lost the film's negative at some point, so it wasn't possible to go more in depth with the analysis. Furthermore, despite help from Joe Zarzynski and others, Sandra was never able to relocate (or was unwilling to reveal) the exact spot where the photo had been taken. Still, further reinforcement came from cryptozoologist and oceanographer Paul LeBlond, who studied the wave patterns in the image and determined the mysterious creature had been sizable, and it seemed to be generating its own wake. Despite the wrangling, between 1981 and 1982, Sandra Mansi's photo was published in several prestigious publications, including the *New York Times, Time Magazine,* and *Life Magazine.*

Noted skeptic Ben Radford has made the argument that what Sandra Mansi photographed that July day was merely debris – essentially an unusually shaped, rotting, dead tree with a curved root sticking out of the water – that had possibly become animate when its state of decomposition had spawned gasses that propelled it momentarily to the surface. However, through the years Mansi would repeatedly state in media interviews that the thing had definitely been alive and had even twisted its head from side to side as she watched. Up until her death in 2018, Sandra Mansi never wavered from her position that she'd managed to photograph an extraordinary unknown animal in Lake Champlain that day.

A PROTECTED SPECIES

NO COMMUNITY HAS WORKED HARDER TO PROTECT CHAMP'S legacy than the hamlet of Port Henry, New York, which is located on the southwestern shore of the lake, overlooking Bulwagga Bay. Not only does the town display a famous billboard that lists many of the notable sightings in the area, but inspired by the work of researcher Joe Zarzynski, Port Henry's board of trustees drafted and passed a resolution protecting the monster on October 06th, 1980. The law prohibits anyone from harassing or harming the creature. Not to be outdone, Vermont House Representative Millie Small convinced her state's government to pass a similar resolution in 1982. Finally, in 1983 the State of New York passed yet even another law advocating for Champ's safety and well-being. Summarily, you would be hard pressed to find another legendary beast anywhere in the world that enjoys more legal protection than the Lake Champlain Monster.

CHARLOTTE FERRY INCIDENT
IN AUGUST OF 1999, THE REGULAR FERRY THAT RUNS ACROSS the lake between the towns of Charlotte, Vermont and Essex, New York was involved in a notable Champ encounter. According to then Captain Jason Alvarez, a thirteen-year veteran of the lake, an irregular, approaching wake caught his attention, since the surface of the water was particularly calm and flat that day. When the strange disturbance got closer, Alvarez could clearly see the five, distinct humps of some unknown animal, totaling twenty feet in length. The creature swam across the ferry's bow about one hundred feet distant and faded from sight. Some of the vessel's passengers also bore witness to the event.

AUDIBLE EVIDENCE
APART FROM SONAR, ACOUSTIC EVIDENCE FOR THE EXISTENCE of any lake monster or sea serpent is virtually nonexistent. But one exception is an underwater recording made at a depth of twenty-four feet in Lake Champlain's Button Bay on the morning of June, 03rd, 2003. The audiotape was made by a team that had been commissioned by the Discovery Channel as part of a program called *Wild Discovery: America's Loch Ness Monster.* Researcher Elizabeth von Muggenthaler, director of Fauna Communications Research Institute, is a bio-acoustician, essentially a scientist who

studies the sounds made by various animals. Along with her colleague Joseph Gregory, von Muggenthaler and her team lowered military-grade hydrophones (underwater microphones) into the lake in order to monitor and record the acoustic environment. Elizabeth, who had grown up in the area, had always harbored a keen interest in the Champ phenomenon, but was understandably skeptical up until that point. In her own words, von Muggenthaler and Gregory looked at each other in "wide-eyed amazement, at the precise moment when they both heard a distinct series of clicking noises that were being picked up by their hydrophone. Almost immediately, Elizabeth recognized the sounds as being some type of echolocation, or bio-sonar, of the type that is specifically employed by cetaceans – dolphins and whales. Because the noises were digitally recorded, von Muggenthaler and her team were able to visually examine the recording with something called a spectral analyzer, and noted the pattern was remarkably similar to that of beluga or killer whales, though unique in some respects. While there is a population of less than a thousand beluga whales endemic to the St. Lawrence estuary, which connects Lake Champlain to the ocean, the nearest these animals are known to come to the lake is at least 277 miles away. Echolocation of this kind is linked to predatory behavior as well as navigation and apparently requires an advanced brain able to quickly decipher the echoes. Further analysis of the recording demonstrated that the sounds being emitted by the unknown source were strongest around the frequency of 35kHz and went as high as 105kHz, over four times above the range of human hearing. Later the same day, von Muggenthaler and her team documented identical echolocations in two other parts of Lake Champlain. While scientific reaction to this evidence has been indifferent, Elizabeth would later make two other attempts (the last one being in 2009) at capturing even more sounds, with the intention of gathering enough data in order to submit a scientific paper on the phenomenon. To date, the paper has not been published.

BODETTE VIDEO

WHILE THE MANSI PHOTO REMAINS A TANTALIZING CLUE THAT something extraordinary may be inhabiting the vast waters of Lake Champlain, there exists an equally important and controversial segment of video footage that most people are not aware of. In July 2005, a local thirty-five-year-old fisherman named Pete Bodette had taken his sixty-four-year-old stepfather Dick Affolter out on his boat in order to do some salmon fishing at the mouth of Vermont's Ausable River. After a while, the two men noticed a log-like object visible from some one hundred feet away. Out of curiosity, the anglers moved in its direction, at which point the thing came to life and swiftly submerged "like a submarine." Around thirty minutes later, the creature, which was at least fifteen feet long, resurfaced, writhing around while displaying a series of small humps on the surface of the water. Pete Bodette, who had fished Champlain his entire life, had absolutely no idea what he was looking at. Fortunately, the men had brought along a video camera and began to film the incident off and on for a span of forty-five minutes. In one remarkable segment, the animal swims by their vessel, just under the surface, before turning and descending out of sight, but not before revealing a fleeting glimpse of a serpentine body the thickness of a human thigh and attached to a small, flat head that seems to display a snout. Whereas there is no dorsal fin visible, some other fins or appendages seem to be apparent just behind the head on its underside. Ultimately, the video was featured on a television episode of *ABC News* and also screened before fisheries scientists at New York's American Museum of Natural History, who apparently were unimpressed (or perhaps couldn't provide an explanation). In subsequent years, the footage has been in the possession of a New Jersey attorney retained by Pete Bodette, who refuses to release the video without significant compensation. As you might expect, this particular situation has created frustration on the part of cryptid investigators.

Whimsical sculpture of Champ in Burlington, Vermont. Photo © Aleksander Petakov

SEARCHERS AND SONAR

APART FROM JOE ZARZYNSKI, THERE HAVE BEEN A HANDFUL OF other passionate and dedicated investigators who have spent considerable time working on the Lake Champlain mystery. Dennis Hall grew up in the area and no doubt, when at a young age he learned his aunt and uncle had claimed to have once seen Champ swim under their boat, it had a profound impact on him. Hall claims to have seen and filmed the monster multiple times himself through the years, starting in 1985. He's advocated a colorful theory that the "creatures" represent a surviving population of a weird, long-necked reptile called Tanystropheus, which lived two hundred million years ago during the lower Triassic period. Hall has even claimed to have once been in possession of a juvenile Champ he found on the shore of the lake, though the invaluable specimen has somehow been misplaced. Scott Mardis has been an active and well-respected Champ investigator since 1994, when he had his only sighting. His longtime partner was Bill Dranginis, an innovative, tech-savvy cryptid researcher who brought his expertise to Lake Champlain starting in 2009. Dranginis designed and built a system of remote, underwater cameras, complete with audio recording capabilities, that could film continuously for up to a month at a time. Tragically, Bill passed away in 2018. Katy Elizabeth, one of the few female investigators tackling the lake monster enigma, has been involved in the quest since 2012. Though she's had a lifelong interest in the Champ puzzle, on her second day of research Katy became a believer when she observed a fifteen-foot, black hump appear and begin to undulate in Button Bay. She's since collected possible video evidence, as well as some intriguing sonar readings. On August 05th, 2019, while piloting her boat back to the dock, Katy's echo sounder recorded two huge, elongate (and seemingly) animate objects at a depth of one hundred sixty-five feet. Other notable Champ hunters include Michael and Diane Esordi, as well as Aleksander Petakov, who has produced an excellent documentary about the phenomenon titled *On the Trail of Champ*.

MEMPHRE
IF SCOTLAND'S LOCH NESS MONSTER HAS A NEXT-DOOR neighbor living in Lake Morar (Morag), then Champ has a kindred cousin in 'Memphre,' the monster of nearby Lake Memphremagog. A mere hundred miles separates these North American bodies of water, though the thirty-two-mile-long, three-hundred-fifty-foot-deep Memphremagog lies mostly in the Canadian province of Quebec, with just a small section descending into the state of Vermont. As with Champ, accounts of Memphre date back to the nineteenth century, and there is also some intriguing photographic evidence. Yet, Memphre has been noticeably more active than Champ in

recent years, based on the number of documented sightings. In fact over two hundred reports have been logged by veteran investigator Jacques Boisvert who refers to himself as a draconologist (one who studies dragons).

One recent encounter involved a resident of the city of Magog, a man by the name of Jean Grenier, who was fishing from his rowboat on the morning of May 18th, 2003. At some point, a vast reflection in the distance just under the water caught Jean's attention. That was abruptly followed by a massive, dark bulge breaking the surface. The only thing Grenier could relate the creature to was some type of whale, at least twenty-five to thirty long. But, Lake Memphremagog is far from any ocean, ruling out the possibility of a stray cetacean. After nearly three minutes, Memphre plunged back into the depths, producing significant waves. In another dramatic incident on June 18th 1989, a fishing guide named Gerard Charland detected an immense, unknown object on his sonar device that seemed to be heading straight for his vessel. "Something that didn't belong there," according to Charland, who knew the lake like the back of his hand. Then to his amazement, the thing surfaced. "It was about the size of a horse. With a tail about twelve to fourteen feet long." To the courageous man's credit, Gerard chased after the animal at full speed when it took off. But as he began to close the gap, the beast sank out of sight. The province of Quebec boasts yet another legendary lake serpent known as Ponik, which is said to inhabit tiny Lake Pohenegamook.

HARRIMAN RESERVOIR INCIDENT

PERHAPS A CHAMP/MEMPHRE-LIKE ANIMAL PAID A VISIT TO southern Vermont on the afternoon of Friday, September 04th, 2015. Lakeside resident Martin Kasindorf, the only person to own a home on the shore of the smallish body of water, noticed his dogs were acting up – aggressively barking at something out in the reservoir around one hundred yards offshore. Looking out, he observed six small humps, about six inches to one foot apart, with water lapping against them. Kasindorf considered the possibility the objects might represent rocks, though he had never noticed them before and knew the reservoir was deep in that particular spot – and either way, he could not understand how rocks were floating on the surface. Martin decided to go and gather his wife for a second opinion, but before he could act, the humps began to move forward and submerge. The next thing he observed was a long, brown object resembling a log "motoring" just below the surface. Kasindorf later studied some videos of swimming otters and other animals, ruling them all out as possible explanations for what he saw.

SOUTH BAY BESSIE

LAKE ERIE, WHICH STRETCHES FROM WESTERN NEW YORK TO Michigan, is bordered to the north by the Canadian province of Ontario. It is the fourth largest of the Great Lakes. Truly though, it is the smallest in terms of water volume, due to it being uniformly shallow. And while the other Great Lakes have produced monster sightings, it is Erie's famous denizen – known as 'South Bay Bessie,' that has earned both a name and reputation. The creature, which displays the serpentine attributes we've covered so far, has frequently been sighted off the port city of Sandusky, Ohio. But, in the early twentieth century, a tugboat crew on the opposite side of the lake claimed they observed the horse-headed creature. As we've discovered, the majority of sea serpent and lake monster accounts from the nineteenth and early twentieth centuries were likely newspaper hoaxes – typically describing sensational features and behaviors and geared toward selling papers. Several such articles mention strange beasties in Lake Erie. In modern times we have more credible encounters, like that of Mary Landoll who was staying at her family's cottage on the shore during July of 1983. Mary observed an unknown animal that displayed a thirty-foot hump as large as an upturned boat, with a head and neck similar to that of a colossal snake. Landoll later stated the specimen looked dark green in color and gave off a strong stench (one of the few aquatic cryptid accounts that describe any type smell). On September 04th, 1990, Bessie was sighted by Navy veteran Harold Bricker and his family while they were out boating. Bricker later described the specimen as being at least thirty feet long, with black skin and a serpentine skull. Yet, his very first impression had been the creature had looked "whale-like."

Bessie doubters have frequently suggested the sightings can be attributed to large sturgeon, prehistoric-looking fish that were once plentiful in the biologically productive lake. However, overfishing in the nineteenth century essentially drove the species to the brink of extinction. Although a 216-pound specimen was caught by an angler in Lake Erie as recently as 1929. In the concluding chapter, I will address the possibility that outsized sturgeon may account for some, but certainly not the majority, of lake monster reports.

CRESSIE

ONE OTHER LAKE SERPENT IN THE GENERAL VICINITY IS THE monster of Crescent Lake, situated in far eastern Canada, on the island of Newfoundland. Because we already have a Nessie, a Chessie, and a South Bay Bessie, it makes perfect sense that this particular creature is named (you guessed it) 'Cressie!' Locals from the adjacent town of Robert's Arm have been encountering the legendary beast since the 1950s and centuries before, the local Beothuk tribe told of a "pond demon." Modern witnesses characterize Cressie as being brown, between twenty to forty feet long and as having a long body and head. "At the beginning, what came to mind was a horse's head. But, it was not as a big… more slim and more pointy," recalled lakeside resident Effie Colbourne, who watched the animal on June 06th, 1995 while looking out her window. She described how its movements created a "deep swell." Effie's sighting lasted a total of fifteen minutes. Other observers have noted the classic "upturned boat" or "snake-like undulation" narratives consistently associated with other lake monster accounts. A popular theory is that Cressie may represent some type of enormous eel that has found its way into the lake from the nearby Atlantic Ocean. But the similarities to its contemporaries – Champ, Ogopogo, Nessie and others – suggests the Crescent Lake Monster is a member of the very same tribe rather than a monstrous eel.

Although a handful of legendary lake serpents have managed to garner worldwide acclaim, there remain dozens of other examples that aren't nearly as well publicized. Some of these will be discussed in the following chapters.

6

Lesser-Known American Lake Monsters

"Men really need sea monsters in their personal oceans. An ocean without its unnamed monsters would be like a completely dreamless sleep." - JOHN STEINBECK

While there are only a handful of North American lakes that can claim a bona fide celebrity monster, there are literally dozens of lakes across the continent where there exists some local tradition or story about a cabalistic creature. To the skeptic, this recurring trope is proof that the situation is indicative of a cultural phenomenon, nothing more. It's essentially a quirk of our age-old beliefs – the notion that large bodies of water conceal giant and often dangerous, unknown beasts. There's little doubt that for centuries these parables have come in quite handy to mothers, who've exploited them as cautionary tales in order to discourage their children from playing too close to the water's edge. So lake monster legends have remained relevant for various reasons.

Still, even as I've been attempting to gradually build a case for the extraordinary possibility that lake monsters may actually represent landlocked, prehistoric whales, it's safe, in my opinion, to disregard accounts that emanate from isolated bodies of water deep in the interior of North America, those that are totally cut off from the sea, or are so far inland as to inspire credulity. A perfect example is tiny Walgren (formerly Alkali) Lake in Nebraska, which dating at least back to the 1920s, has been said to be the home of an immense, brown, alligator-shaped animal with a horn-like projection extending from its forehead. Frequently hyped in local newspapers, the bugaboo apparently displays a voracious appetite, gleefully devouring local livestock. Other miniature lakes and ponds can be scratched off of our monster list for obvious reasons, as can many of the more fanciful portrayals that controvert the standard, sea serpent archetype we've established in the previous chapters. And it's important to recognize the vast majority of lake monster incidents reported in newspaper articles of the 19th and early 20th centuries were almost certainly tabloid-style hoaxes – intended to sell papers, particularly during slow news cycles. Occasionally, the odd hotel or resort owner would also fabricate sightings. Like the opportunistic journalists in that bygone era, innkeepers often found the premise of a lake monster good for business. Nevertheless, there are some intriguing cases worth examining.

Flathead Lake, Montana (Wikimedia Commons)

FLESSIE – THE FLATHEAD LAKE MONSTER

MONTANA'S MASSIVE FLATHEAD LAKE, THE LARGEST NATURAL freshwater body west of the Mississippi, lies over five hundred miles inland from the Pacific Ocean. Yet it's vicariously connected to the sea by way of the mighty Columbia River, the same system that reaches Canada's Lake Okanagan, as well as several other alleged monster lakes in the region. Chinook salmon, lampreys, and sturgeon are among the anadromous species that use the Columbia as a superhighway, periodically migrating between the Pacific and intertwined rivers and lakes. It's not beyond the realm of possibility that at some time in the distant past, other marine animals may have traveled this route too. The lake itself is named after the local Flathead tribe, and its lower half lies on their reservation. Some twenty-seven miles long by fifteen and a half miles wide, this inland sea is an outdoorsman's paradise, surrounded by majestic forests and mountains. Wildlife is abundant.

Accounts of something unusual in the lake date back to the Kootenai tribe who also inhabit the area. One tale recounts how a band of warriors once waged battle with the water dragon, showering its impenetrable flesh with arrows. Another speaks of a titanic underwater beast with giant antlers, which it uses to break through the ice during winter. But the first notable sighting by white men occurred in 1889. Captain James Kern was piloting his ferry-steamer across the lake and spotted an object that he at first took to be a twenty-foot log, and then when it began to move, a whale swimming at the surface. Evidently, one hundred passengers and crew also bore witness to the spectacle, and one bystander even got out his rifle and took a shot at the thing!

Later during the 1950s, an idea was hatched to bolster the local tourist industry by sponsoring a fishing competition. The contest was in earnest an invitation to anglers far and wide to travel to the lake and attempt to catch the lake's monster (which at that time was largely believed to be an outsized fish). The hefty sum of one thousand dollars was put up as a prize to whoever could haul in the big one. However, controversy ensued on May 28th, 1955 when Polson resident C. Leslie Griffith pulled into town with a seven-and-a-half foot-long, 181-pound white sturgeon he claimed to have reeled in near Wild Horse Island. It wasn't simply the fact that no one had ever caught such a big sturgeon in Flathead Lake before; it was that no one had ever caught any sturgeon in Flathead Lake – period. Griffith was reluctantly paid the reward, though many had their doubts and considered the crafty fisherman had actually hooked the specimen in one of the nearby rivers or even to the north in Canada's Kootenay Lake, which has been known to produce some giants. A taxidermy mount of the disputed monster sturgeon can still be seen at the Flathead-Historical Museum in Polson.

An offbeat story from 1960 involves the Ziegler family, who resided in a home on the shore. On one particular day, the clan had noticed there appeared to be large waves splashing up against their dock, despite the fact the lake's surface was relatively calm at the time. When they went out to investigate, the family was astounded to see a huge, horse-headed creature seemingly scratching its elongate body on the pier's wooden pilings!

I RECENTLY INTERVIEWED A GENTLEMAN WHO ENCOUNTERED A massive and mysterious animal at Flathead Lake as recently as late September/early October of 1993 or 1994. His account has never been published before. Travis Reum often goes by the alias 'Raven.' He is a gifted blacksmith and artist who grew up in the area. According to Travis, on that particular morning he was out fishing for Mackinaw trout near a spot called Johnson's Point. The surface of the lake was "like glass… flat and smooth." At one juncture, Travis noticed there was an unusual rippling effect occurring near where he was casting his line. He thought it odd, because he had already caught a few trout, and now suddenly, he was not getting any action – and also couldn't understand what was creating the disturbance. All at once, he caught a brief glimpse of something large breaking the surface about thirty yards out. A moment later the thing resurfaced. Travis explains, "So I kept watching. Then, as I cast out I see the biggest dorsal fin I've ever seen in my life, come up out of the water, as this giant monster comes over the rock shelf. Its fin looked like a shark fin and as it came up, the creature's back came out of the water too, showing a dark almost black-green, scaly skin. I couldn't believe what I was seeing! The water was rippling where it swam in the shallows and it circled the little bay and then headed back out to the deeper water, again showing the dorsal fin as it swam over the rocks. This fin looked to be at least a foot and a half high and a good twenty inches long. It was all one color of dark green with no other markings." Needless to say, Travis quickly hightailed it out of the area and never fished that part of the lake again.

On the afternoon of July 28th, 2005, Judge Jim Manley and his wife Julia had anchored a pontoon boat out in Big Arm Bay on the west side of the lake. The weather was pleasant, and there were no other vessels in sight. The couple had been enjoying some light swimming, but climbed back onto the craft and realized its battery had gone dead. So the unwilling castaways arranged for a rescue from their daughter, who also owned a boat. As they were sitting and waiting, the Manleys both noticed a loud, rhythmic splashing/slapping sound. Looking out, the couple was apparently amazed to see a serpentine animal, some twenty-four in length, undulating on the surface of the water, with two to four humps continually rising upward, as it slowly swam away. Jim began to shout, "That's it! That's it!" realizing he and Julia were now gazing at the lake's legendary beast. By the time their daughter arrived on the scene, Flessie had faded from sight.

❧

RECENTLY, I BEGAN A CORRESPONDENCE WITH LANEY HANZEL, a retired fisheries biologist who spent thirty years on Lake Flathead managing populations of various species, including bull trout and kokanee salmon. Laney particularly enjoyed sharing with me the saga revolving around the 'Mysis shrimp invasion of 1968.' He correctly pointed out in the case of an alleged lake monster, there are ecological considerations one must factor into any theory. In this case, thousands of the tiny shrimp had been dumped in the lake in order to feed the salmon stock, but instead devoured all of the plankton, virtually eradicating the salmon population as a result – and thereby eliminating a potential food source for anything larger. Hanzel tells me he never saw the monster himself in his three decades on the water, nor had he recorded any unexplainable targets on his high-tech sonar rig. Though on more than one occasion, he'd brought up fishing nets that had huge holes ripped in them by something large and unseen. Still and all, through the years Laney had heard enough credible accounts from other locals that he had become convinced there must be some truth to the monster stories.

Then shortly before he was contemplating retirement, Laney was contacted by "self-proclaimed" cryptozoologist Dr. Eugene Lepeshkin of Vermont. Lepeshkin asked Hanzel to begin cataloguing all of the Flessie reports, which he achieved with the help of a local newspaper editor named Paul Fugleburg, who for years had been keeping an archive of the monster's appearances. To date, Laney has compiled 113 sightings, which appear to indicate two completely different types of creatures. About 25 percent of the accounts seem to describe a gigantic fish, up top ten feet long that resembles a sturgeon, while the other 75 percent clearly portray a black, serpentine animal twenty to forty feet long, with "steely black eyes." The most active year was 1993, when no less than thirteen encounters allegedly took place. Laney has even created an impressive map that shows the locations where the monster has reportedly been seen. The most recent report occurred in 2017.

TAHOE TESSIE

THE SECOND DEEPEST LAKE IN NORTH AMERICA AT AN astonishing 1645 feet, Lake Tahoe is a tremendous body of alpine freshwater nestled in the Sierra Mountain Range, with amazingly good water clarity, which drives a thriving tourist industry. It is also apparently home to a beloved monster known as 'Tessie.'

According to an article that appeared in California's *Sacramento Bee* newspaper on July 04th, 1984, two women had sighted a huge animal in Lake Tahoe on the Nevada border the previous month. Patsy McKay and Diana Stravarakis of Tahoe City were hiking a trail just above the lake when they looked down and noticed an object just below the surface that they at first took to be a sinking boat around seventeen feet long. "Then it just kind of emerged like a submarine," McKay told the reporter from the *Bee*. "It went in a circle and then submerged briefly and came back up again." The news story also went on to describe a 1983 encounter that involved two Reno police officers, Kris Beebe and Jerry Jones, who were water skiing when a massive, eighteen to thirty-foot creature glided by their boat like a torpedo, just under the surface. And credit to investigator John Kirk for digging up a 1991 observation made by an eyewitness named Andrew Navarro, who was out pleasure boating at noon one day. "The first thing that I saw was water shooting out of the lake, like when a whale blows water out of its blow hole." Navarro told Kirk, "This was followed by the hump of a brown creature which came out of the water. It moved around in a circle for a while and then it was gone. The movement of the creature was more up and down, not side to side like a snake... My first thought that it was a whale since the creature had to be huge from the size of the hump." One humorous anecdote involves famed diver Jacques Cousteau. It is said that back in the mid 1970s, Cousteau dove deep into Lake Tahoe, and when he surfaced, he exclaimed, "The world is not ready for what is down there!" As it turns out, this particular incident appears to be a lighthearted rumor, perhaps started by Cousteau's son.

While it's not connected to the ocean, Lake Tahoe flows out into the 121-mile Truckee River, which in turn drains into Nevada's Pyramid Lake. And according to local traditions, a "serpent devil" is said to haunt that particular waterway too. During the 1960s, a Nevada state assemblyman even proposed that it become "illegal to annoy the sea monster of Pyramid Lake." Still, the stories are vague. However, I recently spoke to a woman named Christine Trice who, along with her husband, claims to have had an unnerving encounter on Pyramid Lake back in 1994. Evidently, the couple was out Jet Skiing on a windy day when Christine's husband's craft took on some excess water and stalled out. As the couple were floating side by side in the middle of the lake, a gigantic creature suddenly surfaced in the ten-foot gap that separated them. Christine was shocked. She related to me the animal was greenish-brown in color and scaly-looking, though she never did see the thing's head or tail, only its long back, which rolled forward for what seemed like an eternity. Working frantically, Christine's husband finally managed to restart his Jet Ski and the startled couple hightailed it back to shore.

If not a monster, one has to truly wonder what people are seeing in Nevada's Lakes, since the location is a fishing haven and the trophies mostly involve large trout, some of which can grow to lengths of five feet and weights up to thirty pounds. Although, when asked if there were any giant fish inhabiting Lake Tahoe, one local guide quipped that the size of an uncaught fish is relative to the amount of beer one has consumed!

Portrayal of the seminal Lake Iliamna Monster sighting © Bill Rebsamen

ILLIE – THE LAKE ILIAMNA MONSTER

THE SECOND LARGEST LAKE COMPLETELY ENCOMPASSED WITHIN the United States is Alaska's vast Lake Iliamna. Seventy-seven miles long by twenty-two miles wide, it has a medial depth of 144 feet, yet there are deep troughs that may go down to one thousand feet. Iliamna is also isolated. There are no connecting roads, so it can only be reached via floatplane or watercraft, and there are barely a hundred people living in the entire region. Because it is connected to the nearby ocean by the relatively short and wide Kvichak River, the lake boasts some unexpected species, such as a group of four hundred harbor seals *(Phoca vitulina)* and even periodically a stray beluga *(Delphinapterus leucas)* or killer whale *(Orcinus orca)*. There also appears to be a mysterious inhabitant known as Jig-ik-nak, which has been known to the indigenous Aleut tribe for centuries. As the result of modern sightings by white men, the monster has also been nicknamed 'Illie.' Native myths paint the creature as being menacing – known to capsize canoes and devour braves, as well as the occasional swimming caribou. Women and children are warned not to gaze into the water for too long, and also to avoid wearing the color red, since both actions are said to agitate the beast.

However, setting aside any dramatic parables, there have been some compelling sightings in modern times. In autumn of 1941, for example, a forestry worker named Carlos Carson was piloting his floatplane above the mouth of Talarik Creek and noticed what at first appeared to be several floating logs. When the objects suddenly submerged, Carson and his co-passenger realized the "logs" had actually been living creatures of some kind. The very next year, a well-known bush pilot and guide named Babe Aylesworth was transporting local sportsman Bill Hammersley from the village of Naknek to the city of Anchorage. As they overflew the lake, Aylesworth suddenly pointed and exclaimed, "My God! What big fish!" Glancing down, Hammersley saw what Babe was talking about – several torpedo-shaped animals with "blunt heads" swimming just under the surface. Making another pass, Aylesworth estimated the creatures were at least ten feet long – perhaps even twenty feet – and comparable in size to the pontoons on his plane. Watching the critters swim away, the men ruled out whales, sturgeon, or salmon as possibilities. When word of their encounter got out, a Coast Guard pilot named Larry Rost also came forward and stated he, too, had observed one of the apparent monsters while overflying the lake.

During the 1950s, efforts were made to solve the riddle, including a 1959 expedition organized by Texas oil millionaire (and Yeti/Bigfoot hunter) Tom Slick. A reward of one thousand dollars was offered to anyone who could bring in one of the creatures. In the years that followed, there would be many more Illie reports, mostly stemming from floatplane and helicopter pilots who had overflown the vast, freshwater sea. Generally, the descriptions were of gray or black aquatic animals ranging between ten to thirty feet in length. For decades, cryptozoologists have speculated about the identity of the Lake Iliamna monsters, often suggesting the accounts might indicate a subspecies of isolated monster sturgeon, or perhaps misidentifications of beluga whales. One intriguing hypothesis holds that in rare instances, large, Pacific sleeper sharks *(Somniosis pacificus)* might enter the lake. These enormous elasmobranchs can grow up to twenty-three feet in length, and there is a suggestion that, like bull sharks, Pacific sleeper sharks may be able to adapt to freshwater, through a process known as osmoregulation.

An intriguing video captured in 2009 at Nushagak Bay (near the mouth of the connected Kvichak River) adds to the Iliamna mystery. A local man named Kelly Nash was fishing with his two sons when they observed several large creatures swimming by that they were unable to identify – and which were being pursued by beluga whales! Thinking quickly, Nash grabbed a video camera that they'd brought along and began to document the alarming incident. The animals being chased seemed to be serpentine, around twenty feet in length and undulating rapidly through the surf. Their backsides had a serrated look to them. At one point, one of the things apparently even lifted its camel-like head out of the water! The resulting footage was ultimately screened for a couple of investigators – John Kirk and Paul LeBlond, who were duly impressed with the video, particularly because they could clearly discern the beast's head could not be related to any known marine species. The footage was ultimately purchased by a television company that decided to only air a few, fleeting glimpses of the cryptids and failed to include the remarkable 'head shot.' When John Kirk made an inquiry, he discovered that tragically, the Nash family had accidentally recorded over the most significant section of the video before selling its rights!

Regardless, the undulating forms seen in the Nash footage are surprisingly reminiscent of the worldwide accounts of sea serpents, as well as the Loch Ness Monster. Furthermore, Alaskan Inuit legends pay tribute to a great, snake-like monster called the Pal-Rai-Yuk or Tizheruk that is said to frequent the coastline. All things considered, the possibility that Lake Iliamna's famed beasts constitute yet another example of a Nessie-like creature cannot be discounted.

Is the White River Monster a giant alligator gar? (Public Domain)

WHITEY – THE WHITE RIVER MONSTER

THE FATHER OF CRYPTOZOOLOGY, DR. BERNARD HEUVELMANS, once wrote that one could hardly hope for better circumstantial evidence for the existence of lake monsters than the fact that virtually all of the bodies said to host Nessie-like creatures lie along similar lines of latitude in the northern hemisphere – between 30° and 60° north (with the heaviest concentration between 40° and 50°). I believe we've already established that Loch Ness and its counterparts are all cold, deep, glacial lakes, largely inhabited by salmonid type fishes and ultimately connected to the ocean. There are some thought-provoking exceptions to this rule, however: one being the highly publicized monster of Arkansas' White River, affectionately known to the locals as 'Whitey.'

There are rumors the indigenous Quapaw tribe long lived in fear of "a floating island" that would suddenly appear in the river and then mysteriously vanish. But this convenient story may have been contrived after the fact – such is the sensational hype surrounding the beloved creature that is said to haunt the deep river near the community of Newport. There is even one popular legend about Whitey ramming a Confederate steamboat that was transporting gold during the Civil War, causing the lost treasure to sink to the bottom of the White River. This also sounds like a fabulous yarn – something for the locals to ponder as they congregate in the barbershops and diners on slow days. In earnest, the first true sighting was said to have taken place back in 1915.

Yet, the White River Monster truly burst onto the scene in a big way in 1937, due in large part to a local landowner named Bramblett Bateman. On the first day of July of that year, the farmer, who owned a swath of land overlooking the river just south of town, let it be known that he and his workers had been seeing some huge, unknown animal swimming around and scaring away the fish. Bateman described the thing as being about twelve feet long and as possessing a wide, gray-colored body, as well as "a face like a catfish." When the residents of Newport began to snicker behind his back, Bramblett took the weighty step of signing a sworn affidavit before a judge, attesting to the veracity of his claims. Although when he inquired (with officials) over whether or not he could use dynamite on the monster, his request was emphatically rejected.

It wasn't long before word got out about the curious visitor, and Bramblett capitalized on the situation by charging an admission price to come onto his property. Up to three hundred hopeful monster spotters showed up all at once. A circus-like atmosphere ensued, with vendors setting up hamburger and soda stands. A wooden dancefloor was even constructed on-site when a bluegrass band was hired on. It should come as no surprise that Whitey skipped out on his own coming-out party entirely and wasn't seen. Then, the mood took a turn for the serious when a group of local men began to construct a gigantic, forty foot net in order to try and catch the creature. One resident even showed up with a Tommy gun! An ex-Navy diver from Memphis named Charles Brown soon arrived on the scene with an eight-foot harpoon. Brown's one-man, five-day expedition under the water made national news. Still and all, the elusive beast was nowhere to be found. Events around Newport quieted down for a spell. It would be thirty-four years before the monster would return.

In June of 1971, Newport's *Daily Independent* newspaper received a call from a well-respected, local businessman who wanted to remain anonymous. The man claimed he had just watched an animal as "large as a boxcar" splashing around in the White River. The critter's mostly smooth skin had apparently looked gray and seemed to be peeling away in spots. Shortly after the paper published that account, another resident named Ernest Denks came forward, stating he, too, had seen the thing while out boating. "It looked like something that came from the ocean," Denks stated. "It was gray, real long and had a pointed bone protruding from its forehead." Ernest didn't stick around long enough to study the

creature, which he estimated to weigh a half a ton. On June 28th, a famous photo was taken, alleging to show Whitey rising to the surface. The Polaroid picture was snapped by fisherman Cloyce Warren, who was in a boat with two buddies less than two hundred feet away from the monster. Warren was scared to death: "That thing looked like something prehistoric. It was over 30' long. Its tail was thrashing." Unfortunately, the resulting image leaves a lot to be desired. Though blurry, it appears to show a long, serrated tail with something of a blob in front. It's hard to ignore the fact that what appear to be spikey scutes on the tail are reminiscent of those found on an American alligator *(Alligator mississippiensis)*. So perhaps some encounters with large gators have gotten into the mix somehow.

By the 05th of July, Sheriff Ralph Henderson had been notified some strange tracks had been found on Towhead Island. The inquisitive law enforcement officer went out to investigate and saw the prints for himself – fourteen inches long by eight inches wide, featuring three toes with long claws and a spur extending backward at an angle – impressed into the sand. In addition, a large swath of vegetation had been flattened by something heavy. According to White River Monster historian Jason Mansfield, it turns out there's a strong possibility the tracks had been hoaxed by local teenagers who were looking to take advantage of the monster hysteria. Still, there was another remarkable incident involving two men who alleged Whitey had actually lifted their boat up out of the water! On July 23rd, Ollie Ritcherson and thirteen-year-old Joey Dupree were angling for catfish, when without warning, their vessel raised up above the surface of the water, turned ninety degrees and then dropped back down with a distinct thud. Though they didn't see what had caused their craft to elevate, it was suggested Whitey was probably the culprit.

However, I recently corresponded with Joey Dupree himself, who expressed to me in retrospect he is firmly convinced the boat had probably been lifted by a huge alligator gar *(Atractosteus spaptula)*, and that the media simply ran wild with the story. Not unlike when it reported that a UFO investigator from Oklahoma City arrived in Newport, exclaiming Whitey was an example of an "experimental animal sent to Earth by aliens, much as man would send a monkey into space." In the end, the White River Monster gained in popularity to the point where a law was passed in the Arkansas legislature, essentially creating a "refuge" where Whitey would be protected from harm (similar to the case with the Lake Champlain Monster). The last known sighting of Whitey (and possibly his mate) transpired in early April 1974. Leon Gibbs and Carl Jackson were running fishing lines on two separate days when they noticed two large swimming animals that reminded them of black cows, although the men had both decided the humped creatures could not be cattle, due to the fact that the beasts swam against the current.

Evaluating the reports, one reaches the inescapable conclusion that Whitey does not seem to match with our Nessie-type specimens. So, what does it represent? There are some viable theories that involve known animals. For example, as Joey Dupree suggested, the White River is known to be home to huge alligator gars that can grow to lengths of eight feet and weigh over three hundred pounds. Though quite frightful and prehistoric-looking in appearance, these elongate predators really don't match all of the descriptions. Cryptozoologist and Loch Ness investigator Dr. Roy Mackal suggested perhaps a stray bull elephant seal (*Genus: Mirounga*) had meandered up through the Gulf of Mexico into the Mississippi River and ultimately into the White River. These massive marine mammals can grow to be over twenty feet in length and weigh over eight thousand pounds. In addition, their skin is generally gray and can display the "peeling" described by some witnesses. However, the Gulf of Mexico lies thousands of miles outside of the known range of either of the two known species of elephant seals. Thus, perhaps the most palatable explanation might be that the legend of Whitey was spawned by wandering Florida manatees, also known as sea cows – aquatic mammals that can grow up to fifteen feet in length and weigh over three thousand pounds. As recently as 2006, one wandering individual swam over seven hundred miles up the Mississippi River to the city of Memphis, which lies less than one hundred miles southeast of Newport.

NEUSSIE?
FROM TIME TO TIME I'M APPROACHED BY AN INDIVIDUAL WHO claims to have knowledge of a Nessie-like sighting. Such is the case with a gentleman from North Carolina named Douglas Helvie. In truth, Douglas was speaking on his ex-wife's behalf, since at the time he was driving their truck across a busy, one-lane bridge at the time of the incident and didn't feel it was safe to take a gander for himself. The incident occurred over the mouth of the Neuse River,

which flows into the Atlantic Ocean (less than one hundred miles south of Chesapeake Bay). All at once, Doug's normally "very down to Earth" ex-wife began to excitedly point and shout, "Do you see it? Do you see it?" at which point he attempted to take a quick glance but couldn't tell what she was pointing at. "One of those Ogopogo things... like you know, in Loch Ness!" she continued. Once the couple had crossed the bridge, Doug was able to elicit more details. His wife had watched the animal for a span of two to three minutes, during which time its sizable head and neck repeatedly broke the surface of the water. In her opinion, the monster had a prehistoric-look to it. Interestingly, according to local author John Hairr, the Tuscarora people of the region have legends about lake creatures, and evidently a sea serpent has also been sighted off the coast near Wilmington, North Carolina. What is that monster's name? You guessed it – 'Willie!'

7

Lake Monsters around the World

"Man cannot discover new oceans unless he has the courage to lose sight of the shore." - ANDRE GIDE

Similar worldwide traditions of lake monsters add yet another degree of evidence supporting the potential reality of aquatic cryptids like Nessie. Once again, the eyewitness descriptions remain consistently uniform. And as in both Scotland and North America, the types of lakes are similar too: generally deep and cold (glacial) and positioned along the same isothermal lines of the northern hemisphere.

ICELAND'S LAGERFLJST WORM
THE FRIGID ISLAND NATION OF ICELAND IS RIFE WITH LEGENDS of sea monsters (known collectively as Skrimsli) – from mermaids and scaly, black, hippo-sized beasts to a shaggy, sheep-like amphibian known as the Fjørulabbi ("Shore Laddie"). The most famous creature though, is said to dwell in the chilly waters of Lake Lagerfljøt. There is even a popular and fanciful fable that explains the genesis of the so-called 'Worm.' Long ago it seems a young girl was given a gold ring by her mother, who instructed her to lay the

precious heirloom on top of a worm within a large wooden box. The child's mother explained that in this way, the size of the ring would grow, increasing its value. Yet after a great deal of time had passed, the anxious girl opened the box prematurely, only to discover that while the ring had remained the same size, the worm had grown into a veritable monster! Horrified, the girl threw the box, worm, ring and all into Lake Lagerfljøt – where the immense creepy-crawly dwells to this very day.

However, a surprisingly pragmatic account of the monster worm dates all the way back to the mid-fourteenth century, where it was chronicled in a historical text known as the *Landnamabok*. According to the old script, one particular summer there appeared "A wonderful thing in the Lagerfljøt which is believed to have been a living animal. At times it seems like a great island, and at others, there appeared humps…. with water between them." This sounds surprisingly like most of the modern sightings of lake monsters we've covered so far. Writing about his personal explorations of Iceland in 1861, Rev. Sabine Baring Gould mentioned an esteemed doctor and academic he knew who'd once stumbled upon the remains of a huge, unknown creature on the shore of Lagerfljøt. The physician had determined it wasn't the carcass of a whale. Regrettably though, there is no record of a sample being preserved.

Locals who've grown up in the sparsely populated region seem to have generally accepted there is something queer living in the lake. And almost every resident knows someone who claims to have seen the thing over the years. During the 1960s the monster was allegedly spotted by some forestry agents, and two decades later, utility workers replacing telephone cables beneath the depths found that the housings had been mangled by some unknown force. There is also an account from 1998 that mentions a local teacher, as well as some of her students, who watched the monster cavorting in the lake through the schoolhouse window.

Then in 2012, a video captured by a sheep farmer living on the lake's edge was uploaded to YouTube, where it has subsequently been viewed over five million times! The footage, which was featured by media outlets around the globe, appears to show a sizable serpentine form swimming on the surface of the icy water. At first glance, the image is somewhat startling. But upon closer inspection, the object is not actually making any forward progress. A swift current flowing against it merely makes it appear that way. Perhaps most revealing is the fact that the anomaly exhibits a rather rigid, side-to-side motion. For this reason, it has been suggested that, rather than a living creature, the entity represents a long mass of debris, such as a frozen fishing net, swaying in the current. Yet perhaps not wanting to miss out on any potential tourism opportunities, in 2014 a panel commissioned by the Icelandic government ruled the controversial video was genuine and the mystery of the Lagerfljøt Worm bears further investigation.

SCANDINAVIA – STORSJSODJURET & SELMA

DUBBED SWEDEN'S LOCH NESS MONSTER, ACCOUNTS OF A creature inhabiting that country's fifty-fathom deep Lake Storsjøn supposedly date back almost four centuries, to the year 1635 AD. There's also a thousand-year-old standing runestone (tablet), discovered on one of the lake's islands, which features an engraving that depicts a dragon-like serpent. Known locally as the Storsjøodjuret, the beast is said to fit the typical sea serpent mold in some ways – up to forty feet long, often resembling a log or upturned boat from a distance and undulating vertically when it swims. Yet there are some unique traits. Older reports portray the animal as having a broad, dog-like (rather than horse-like) head, and perhaps most intriguing, as having white, tube-like structures on either side of its neck, just below the jawline. These odd appendages have been likened to everything from a mane, to floppy ears, to small fins. Another unexpected detail is multiple witnesses have described the Storsjøodjuret as having a reddish or cinnamon color, as opposed to the typical gray or black hue associated with other lake serpents, and also as sometimes having spots.

Decades before Nessie's rise to fame, the Lake Storsjøn monster was Europe's most celebrated lake mystery. In 1898, a local marine biologist and schoolteacher named Dr. Peter Olsson took an interest in the riddle and gathered twenty-two reports from various citizens of veracious reputation. An official hunt was even organized, financed by a wealthy, local matriarch (with the full-hearted support of Sweden's King Oskar II). It involved an experienced whaler armed with custom harpoons, as well as a giant steel trap baited with a dead pig. Needless to say, the venture was unsuccessful. Ironically, nearly a century later in 1986, the Swedish county of Jåmtland passed a declaration that protected Storsjøodjuret from being killed or captured (following in the footsteps of America's White River and Lake Champlain Monsters). The law was repealed in 2005 for some reason.

A dramatic encounter with 'Storsjie' unfolded in 1973. Fisheries officer Ragnar Bjørks was on the lake in a twelve-foot rowboat when allegedly the forked tail of a colossal creature suddenly broke the surface of the water adjacent to his location. Bjørks later recounted how, when its back also rose up, the torpedo-shaped beast seemed to be at least eighteen feet in length. Though mostly displaying a brownish-gray color on top, the thing's underside was a lighter, yellow color. Its skin appeared smooth and "whale-like." Perhaps borrowing from his fisheries department training, Ragnar raised his oar and began to vigorously beat the poor animal's body. According to the account, a defensive whack of the monster's tail reverberated through the frantic game warden's craft, knocking him backward.

At 12:21 p.m. on Thursday, August 28th, 2008, one of six stationary video cameras that had been deployed at various points around Lake Storsjøn captured "the movements of a live being." The project was the brainchild of Gunnar Nilsson, head of the local shopkeeper's association in the lakeside locality of Svenstavik.

Although it's difficult to judge size and scale, the underwater video clearly shows a long, twisting, snake-like organism with small, flat head attached. Because the sentry cameras had been utilizing thermal heat-signature technology, the reddish aura emanating from the mystery creature seems to imply that – whatever the thing was – it had been warm-blooded.

Norway's relatively small Lake Seljord has a storied lake monster legend too. As we discovered in Chapter Three, Norway's importance with regard to the enduring sea serpent saga cannot be understated. If there is any nation that rightfully should have a resident lake beast, it feels as though the Land of the Midnight Sun ought to be the place. Affectionately called 'Selma' by residents of the tiny municipality of Seljord, the animal was allegedly first encountered back in 1750 and has been observed by a number of credible residents in the intervening centuries. Selma is generally described as being between fifteen to forty feet long and snake-like, with smooth, dark skin and a horse-like head.

Beginning in 1977, Swedish journalist turned cryptozoologist Jan Ove-Sundberg organized a number of expeditions at Lake Seljord in order to search for evidence of the enigmatic animal. On his very first trip out onto the water, Jan claimed his state-of-the-art Simrad sonar unit almost immediately detected a large, animate object coming directly at his boat! Fortunately, his flimsy craft was not upended and he managed to record other sizable contacts as the evening wore on. This was enough to convince Jan there was a genuine mystery afoot. Under the umbrella of his organization *GUST (Global Underwater Search Team)*, Jan would utilize numerous volunteers and gadgets for over two decades, employing sonar, scuba divers, remote-controlled mini submarines, hydrophones, and even ultra-light airplanes flying overhead the lake. Sadly, Ove-Sundberg passed away in 2011, never fulfilling his dream of proving Selma's existence.

On November 11th, 2010, helicopter pilots Even Birkeland and Eddy Dale were surveying Lake Seljord from the air when they spotted two huge objects cutting through the water and creating a massive disturbance. The stunned observers began to take a series of pictures, which would ultimately make headlines in the local newspapers. According to Birkeland and Dale, who had been flying over the lake for years, they had never seen anything quite like the two anomalies, which appeared to be cruising just below the surface of the water, pushing up a tremendous wash and causing V-shaped wakes. As the helicopter approached, the creatures changed direction and submerged. Lake monsters have apparently been reported from over fifty other Scandinavian lakes, including Rommen and Mjøsa.

Portrayal of a 1960 lake monster sighting in Lough Ree, Ireland © Bill Rebsamen

IRELAND, ENGLAND & WALES

AS YOU MIGHT EXPECT, THERE ARE NUMEROUS REPORTS OF creatures like Scotland's Loch Ness Monster stemming from lakes in the adjacent countries of Ireland, England, and Wales. The Emerald Isle, for example, features aged traditions of mythical water beasts called 'Peistes' or 'Horse Eels,' not that dissimilar from Scotland's so-called Water Kelpies. There are older stories of these "wurms," described as growing up to thirty feet in length and looking essentially eel-like, but with a mane of hair (or perhaps a fin) running down the neck and back. They have even been said to slither across the land from lake to lake – occasionally getting stuck in a culvert or under a bridge. In June 1954, four people fishing were allegedly horrified by a monstrous Peiste while boating on Lough Fadda, a small body of water nestled within the peat bogs of Southwest Ireland. Librarian Georgina Carberry recalled the creature had blue-black skin and seemed to display movement throughout its creepy body. When it swam up to their small vessel and opened its mouth wide, Georgina and her friends jumped up, causing the varmint to retreat behind a rocky island. The woman was so upset by the encounter that she had repeated nightmares about the incident. Over a decade later in 1965, Capt. Lionel Leslie of the *Loch Ness Investigation Bureau* arrived on the scene with some explosives, which he discharged in the depths of Lough Fadda in an attempt to drive its mysterious resident to the surface. The experiment was evidently unsuccessful.

In May of 1960, three priests were fishing from a rowboat on Central Ireland's large but shallow Lough Ree when they noticed a great animal about one hundred yards away. The clergymen – Rev. Daniel Murray, Rev. Matthew Burke, and Monsignor Richard Quigly – would later describe the thing as having a pronounced hump with a slender neck and small head sticking out at a thirty-degree angle, about six feet in front. The head appeared to be flat like a python's. In their astonished estimation, the creature had looked much larger than any species that should have been living in the lake. The figure moved steadily away from the priests for the two to three minutes that they studied it, submerging and resurfacing a couple of times before disappearing for good. Due to the obvious veracity of the witnesses, this particular incident was widely publicized at the time.

One of the weirdest lake monster encounters on record occurred on February 22nd, 1968 at Galway County's wee Lough Nahooin, situated on Ireland's western coast. Literally the size of a football field, Nahooin is the last place on Earth where there should be an unknown critter swimming about. Yet when farmer Stephen Coyne went down to the veritable pond that particular evening in order to gather some dried peat, he was scared out of his wits by the sight of an unusual specimen. Stephen's dog began to bark emphatically, causing the monster to swim closer and open its mouth, as if to say, "back off." Eventually, the entire Coyne family was summoned to the scene and all watched the thing in amazement, not really sure what they were looking at. The creature was about twelve feet in length, with slick, black skin, a periscope-like neck and a flat tail. When it dipped its head underwater, the thing's back would bunch up into two coils. At one point, the beastie swam within twenty feet of the shore, and Mrs. Coyne was able to observe two horn-like structures on the animal's head, though she was unable to make out anything resembling eyes. As darkness fell, the baffled family returned home. This was supposedly the last known sighting of the Lough Nahooin nasty.

Lough Muckross lies in the southernmost part of Ireland. It is in fact the nation's deepest lake with a maximum depth of 246 feet. It is also allegedly home to a monster with the glorious name 'Muckie!' An early account that dates back to 1885 describes how a young lass had waded out into the water and noticed a splashing noise. When she peered out, the girl evidently saw a frightening creature with large eyes swimming in her direction, causing her to flee. In the years that followed, the mysterious two-humped animal was apparently seen by other locals, as well. Interestingly though, as recently as April of 2003, scientists conducting a sonar survey of Lough Muckross recorded "a deep lurking thing," which appeared to be snake-like and about twenty-seven feet long. Just to the southwest of Muckross lies miniscule Lough Brin, where there supposedly resides a water beast known as 'Bran' – described as a "dreadful wurrum" around fourteen feet in length and with big eyes like Muckie. Bran was last sighted in 1954 by a local farmer named Timothy O'Sullivan. Years earlier in 1940, a twelve-year-old boy claimed he encountered the beast basking on the shore.

Windermere in Northern England is that nation's largest body of water and has a depth of over two hundred feet. It is also the apparent home of a monster the local tourism board has conveniently named 'Bownessie.' Strangely, the first publicized sighting didn't even occur until July 23rd, 2006, when a reporter on holiday, along with his wife, watched and photographed a giant, dark shape with three humps moving through the depths. Later, on February 11th, 2011, a kayaker named Tom Pickles (love the name) snapped a photo of an unidentified object with a lumpy backside that left behind a V-shaped wake. Since there have only been about ten Bownessie reports in total, the matter remains unresolved, although England's Centre for Fortean Zoology has mounted multiple expeditions to Lake Windermere and theorizes there may be an extremely large and unusual fish involved.

The country of Wales, which is located on the western edge of England, can claim a monster too: 'Teggie.' Just a few miles in length, Teggie's alleged haunt, Lake Bala, runs into the same problem as many of the putative monster lakes throughout Ireland and the United Kingdom. It just seems highly improbable that any giant, undiscovered species could inhabit the tiny tarn. Still, if the accounts of Irish 'horse eels' traveling short distances across land from lake to lake are accurate, it could potentially explain why these ultra-rare cryptids are only seen periodically in smallish bodies of water.

RUSSIA – BROSNYA & THE LABYNKYR DEVIL

ABOUT TWO HUNDRED FIFTY MILES NORTHWEST OF MOSCOW IN far Western Russia lays Lake Brosno, which, according to legend, is said to be home to a lake dragon. Although traditions of the Brosno dragon, or 'Brosnya' date back to the thirteenth century, a Russian family claimed to have seen and photographed the creature as recently as 1996. Still, as accounts of Brosnya are rare, the odds of finding a lake monster appear to improve dramatically five thousand miles to the east.

The vastness of Russian Siberia is difficult to comprehend – literally five million square miles of unforgiving tundra, marshes, forests and mountain ranges and one of the least densely populated areas on Earth. The Republic of Sakha (formerly the Yakutsk Region) of Eastern Siberia, for example, is over one million square miles, almost the size of India, but with only one thousandth the population. It seems if there are still discoveries to be made, Sakha is as good a place as any. And indeed, there are two isolated lakes there that have produced monster accounts. In frigid, eight-mile-long Lake Labynkyr, there is said to dwell a terrible "devil" that has been seen since the nineteenth century and which is believed to devour swimming dogs and deer on occasion. A dozen miles away lays smaller Lake Vorota, which has also produced sightings. According to a 1962 article by a Soviet zoologist named S. K. Klumov, a famed Russian geologist on a survey mission to the region claimed to have spotted a monster in Lake Vorota in July of 1953. The geologist described the thing as being dark gray in color, up to thirty-six feet in length, with a wide body, wide-set eyes, and a high, backswept dorsal fin. This encounter supposedly was the impetus behind an expedition of Soviet scientists who trekked to both lakes during 1963 and 1964. The team claimed to have observed huge, humped objects moving around in Lake Labynkyr on multiple occasions, but failed to come up with definitive proof of a monster. Around that time period, Radio Moscow announced yet another scientist at a remote, volcanic lake in the same region – Lake Khaiyr – watched a bluish-black, dinosaur-like creature crawl out of the water onto the shore and forage on grass! Due to the sensational nature of these stories, we cannot dismiss the possibility these Siberian lake monsters were the result of Soviet propaganda. It's no secret that at the height of the Cold War, the Soviet Union would frequently make dramatic announcements, claiming dubious scientific achievements. Yet events in recent years may add some substance to these accounts.

In 2002, Lyudmila Emeliyanova, an associate professor of biogeography from Moscow University, was conducting a non-monster-related study at Lake Labynkyr when her sonar unit picked up something unexpected. "It was our fourth or fifth day at the lake when our echo sounding device registered a huge object in the water under our boat," Emeliyanova told *The Siberian Times,* "bigger than a fish, bigger than even a school of fish... As a scientist, I cannot offer any explanation of what this object might be." The sonar reading indicated the anomalous contact was close to twenty feet in length, and Lyudmila was subsequently able to record similar targets on other occasions causing her to conclude, "This mysterious and very deep lake still has some secrets to tell us." In March of 2014, a team of Russian divers conducted the deepest "under ice" dive in the chilly waters of Labynkyr, down to a depth of one hundred seventy feet. Interestingly, the water temperature under the massive ice sheet remained above freezing, despite the fact the air temperature around the lake at the time was 50° below zero! For years, it has been hypothesized there is a warm, volcanic vent seeping into the lake and that Labynkyr and Lake Vorota may be connected via deep caverns. One weird detail: *The Siberian Times* reported the divers discovered a giant jawbone and skeleton on the floor of the lake, similar to those of a prehistoric marine reptile known as a pliosaur. But, no photographs or further mention of the find have ever been published.

TURKEY – CANAVAR

LOCATED IN A REGION STEEPED IN HISTORY LIES EASTERN Turkey's immense Lake Van. In reality, this vast, high-altitude body of water could be considered an inland sea, and in fact for centuries was referred to as the "Upper Sea" by the ancient peoples that inhabited the area. Formed around 600,000 years ago by volcanic activity, the lake is some seventy-four miles across and incredibly fathomless in some spots, reaching depths of 1480 feet. The water level has been rising steadily ever since lava flows from a nearby volcano cut off the outflow to surrounding rivers during the last Ice Age. Perhaps most notably, Lake Van is considered a 'soda lake' due to its high salt and calcium carbonate content. Subsequently, the only native species of fish capable of surviving its inhospitable conditions is a type of minnow known as the pearl mullet *(Alburnus tarichi)*. Despite this actuality, the ruins of a primordial castle nestled on a hill beside the lakeshore feature a stone engraving that seems to depict a huge water beast attacking a boat. And in the nearby Armenian Highlands, there are widespread legends that tell of serpentine dragon called the Vishapakar.

Yet the first modern reference to the Lake Van Monster, or 'Canavar' as it is called, can be found in a Turkish newspaper article from 1889, which describes how three men camping along the shore were attacked by a menacing creature from the depths. In the decades that followed, sightings of the creature were not widely publicized, such as when fisherman Hasann Mollaoglu encountered the monster in 1962. Or, when in 1996, several individuals in the town of Celebibagi allegedly observed the remarkable animal struggle to free itself from a marsh on the lake's northern edge. Canavar gained true notoriety in 1997 when a local researcher named Unal Kozak produced a clear, close-up video of a giant, humped thing moving across Lake Van's surface. The strange object seems to resemble the top of an elephant's head, complete with a visible eye and was even shown on *CNN*. The astonishing quality of the footage has led some investigators to conclude it could represent a clever hoax involving something being towed through the water. Yet another video that surfaced was shot by tourist Nesimi Dede and matches the classic vertical undulations associated with other global lake monsters. Just recently in 2017, divers exploring the depths of Lake Van discovered the submerged ruins of a three-thousand-year-old castle from the Urartu civilization. The unexpected find provides hope that Lake Van may harbor other secrets.

JAPAN – ISSIE

JAPAN'S LAKE IKEDA IS KNOWN AS A CALDERA LAKE, MEANING IT is essentially a dormant volcano that filled up with water around six thousand years ago. As such, despite its close proximity, it is not connected to the ocean and depends on rainfall in order to sustain itself. Located on the large, southern island of Kyushu, Lake Ikeda is small in surface area, but tremendously deep. The local tourist industry has both adapted and heavily promoted its resident lake monster, which is known as 'Issie-kun,' or 'Issie.' And while there allegedly have been numerous sightings in recent years, the first known account of the creature only dates back to 1961. The seminal encounter occurred on September 03rd, 1978 when around twenty students playing soccer at a residence beside the lake looked out and saw a series of dark humps undulating through the water. This incident, which lasted about three to five minutes, was widely covered by Japanese newspapers. It goes without saying that the nation that brought us *Godzilla* seems quite proud of its own version of Nessie. Some photographs taken in December of 1978 by a man named Toshiaki Matsuhara appeared to show the classic sea serpent humps and were considered to be further proof something was afoot. Then, in 1991 a video claiming to show Issie swimming along was shot by a witness named Hideaki Tomiyasu, although things seem to have quieted down since then. It's widely known the lake supports a healthy population of decent-sized, Japanese eels or unagi *(Anguilla japonica)*. But sightings of Issie often include length estimates from sixteen feet to up to one hundred feet, so it's highly unlikely eels are behind the reports.

AUSTRALIA'S BUNYIP

UP TO THIS POINT, ALL OF OUR LAKE MONSTER ACCOUNTS HAVE been from the Northern Hemisphere. And as I suggested earlier, this could be interpreted as an example of circumstantial evidence, advocating for the existence of a genuine unknown species. Traditions of these creatures seem much less prevalent in the Southern Hemisphere, but there are some. On the island continent of Australia, for example, there is a celebrated water beast known as the 'Bunyip.' Now, it must be acknowledged descriptions of this animal vary widely, and in fact the name, which comes from the native Aboriginal tongue, essentially translates to "boogeyman," so it's sort of a catchall term for any mystery critter. It's also worth mentioning virtually all of the reports of the Bunyip stem from the nineteenth century, with few modern sightings. A classic example would be the thing sighted in the Eumeralla River in 1848, characterized as a giant, brown varmint with a long neck and kangaroo-like head. A monster that was encountered in Port Phillip Bay south of Melbourne was said to be the size of a bull but possessing a skinny neck and small head, similar to that of an emu. Yet the majority of Bunyip stories from the 1800s seem to describe something else entirely – often depicted as a shaggy, dog-headed animal (curiously, sometimes having "floppy" ears that it uses to propel itself through the water!) This has led to speculation that the Bunyip myth may be largely based on sightings of wayward seals or sea lions that have wandered up rivers and lakes, far from their typical habitat on the coast.

Perhaps Australia's best claim to an aquatic cryptid similar to the Loch Ness Monster is the bewildering creature that has been repeatedly observed in the Hawkesbury River, just north of Sydney. Sightings may date back to the 1940s and even before, but the entity created a mini stir during the 1970s. In one marked incident that occurred during a stormy morning in 1979, a couple living on a houseboat, Mr. and Mrs. George Cayley, felt some unseen force ram into their moored vessel. When the hysterical pair stepped out onto the rainy deck in order to investigate, they reputedly watched a thirty-foot serpentine life form oscillating away from their location in an up and down motion. During March of 2020, Australian researcher Tony Healy mounted an investigation into the phenomenon. Though many locals he spoke to seemed to reject the idea of a monster in the river, Tony did finally track down some fishermen who claimed to have encountered outsized eels in the winding waterway. Female specimens of the native longfin eel *(Anguilla dieffenbachii)* are known to achieve lengths of over nine feet. Others suggested to him large sea turtles might enter the inlet on occasion and that their prehistoric appearance might invoke startled reactions.

Perhaps most intriguingly, a gentleman named Martin Shaw described to Healy how, back in March of 1999 or 2000, he observed a "very strange-looking animal" close to twenty feet long appear near some moored boats on two separate mornings one week apart near the delightfully named community of Mooney Mooney. The creature did not possess a dorsal fin and moved in an odd back-and-forth fashion. Healy also learned that a beast with a horse-like head and multi-humped body had allegedly been seen at a place called Swallow Rock Reach. Tony plans to return for further investigation into the so-called Hawkesbury River Serpent at a later date.

Argentina's Lake Nahuel Huapi (Wikimedia Commons)

ARGENTINA – NAHEULITO

WHEN MOST PEOPLE THINK OF SOUTH AMERICA, THEY probably visualize the mighty Amazon River Basin and its dense, sprawling jungles. Yet there is much more to this breathtaking continent – from the dusty Atacama Desert of Chile, to the fertile, southern grasslands known as the Pampas. And in western Argentina at the south end of the Andes Mountains lies sparsely populated Patagonia – an alpine region replete with glaciers, high pine forests and magnificent fjords. It's a land virtually indistinguishable from locales like Western Canada and Norway, where we've already established a pattern of lake monster activity. Smack dab in the middle of Patagonia lays Lake Nahuel Huapi – six miles across and with an impressive maximum depth of 1,522 feet – and supposedly home to South America's version of the Loch Ness Monster – known as 'Nahuelito.' Legends of a menacing beast in the lake may date back to the indigenous tribes of the area, who told of a humped creature that possessed a cow-like hide called El Cuero (the leathery one). Accounts of this so-called "Patagonian Plesiosaur" gained notoriety the early nineteenth century when a colorful Texan explorer named Martin Sheffield claimed he observed a massive animal with a swan-like neck surface at another remote montane lake.

    This incident was the inspiration for an expedition launched by the Buenos Aires Zoo in March of 1922, which ultimately failed to turn up any evidence, despite deploying dynamite charges under the water. Regarding Nahuelito, reports remained sporadic until later decades. On a 1997 episode of the popular TV show *Sightings* titled "Monster of the Andes," an eyewitness named Tessie Campbell claimed she was working at a lakeside resort when one of her colleagues pointed out something unusual moving out in the water. Campbell described the object as being "a very big animal… with several humps like a whale." Yet another eyewitness named Hector Ulacia described how the surface of Nahuel Huapi had been smooth as glass when he noted a dramatic disturbance out of which rose something that was "black, very dark and moving." Years ago, quite by chance I met a gentleman from Argentina who claimed to have sighted Nahuelito. But sadly, he never responded to my emails when I attempted to follow up for more information. Quite recently in January of 2020, the Argentinian press ran news stories about two young men who captured distant video of an anomaly on the lake, temporarily reviving interest in the creature. However, the wake appears to be nothing more than an unusual standing wave known as a seiche.

❊ 8 ❊

Some Confounding Carcasses

"There is nothing in the world more stubborn than a corpse. You can hit it. You can knock it to pieces. But, you cannot convince it." - ALEXANDER HERZEN

One might reasonably ask: If the Earth's waterways truly are populated by a species of giant, unknown creatures, wouldn't their remains surface or wash up on the shore from time to time? Certainly, the sea is known to give up some of its nethermost secrets on occasion. In this chapter, we will explore instances where controversial carcasses of marine animals have caused a stir – at least momentarily being considered as definitive proof of aquatic monsters.

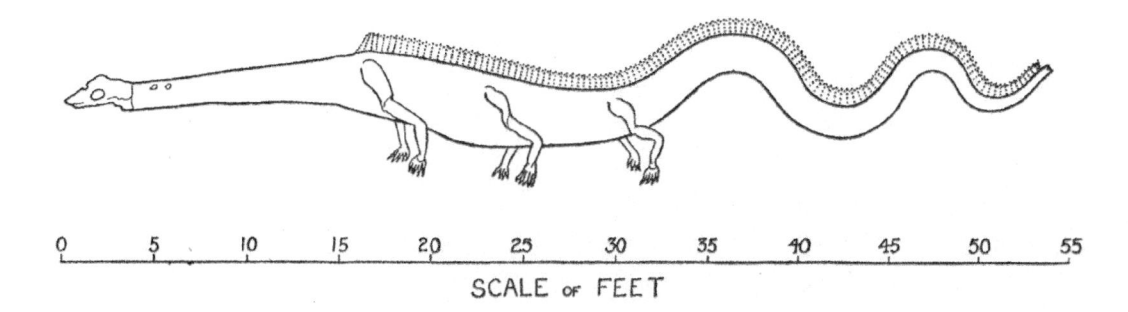

Vintage sketch of the Stronsay Beast, based on witness descriptions (Public Domain)

THE STRONSAY BEAST

LESS THAN TWO HUNDRED MILES NORTH OF LOCH NESS LAY Scotland's viridescent Orkney Islands. Stronsay, one of those small isles, was the sight of a sensational stranding that occurred back in the autumn of 1808. Events began to unfold on September 26th, when farmer turned fisherman John Peace noticed a flurry of seabirds scavenging the carcass of a large animal in the waters about a quarter mile off Rothiesholm Head. Curious, Peace took his boat out to investigate and was bewildered to discover it hadn't been the remains of a whale, as he had assumed, but rather a long, serpentine thing with leg-like appendages that were quite unlike fins or flippers. Peace's inquiry was witnessed from the shore by landowner George Sherar – and when a blowy squall ultimately deposited the oddity onshore several days later, Sherar, Peace and also a local carpenter named Thomas Fotheringhame all examined the corpse independently, making thoughtful observations and taking careful measurements.

The consensus opinion was that the beast was almost exactly fifty-five feet long and about five feet thick, with gray skin, a smallish skull about the size of a seal, a neck between ten to fifteen feet long, an equally long tail with no fin, and six slender limbs that seemed to have toes. In addition, there appeared to be a mane consisting of luminescent, silvery-yellow filaments that were about fourteen inches long running the length of the backside. Fortunately, Sherar also made a crude sketch of the animal, and most importantly, saved some of its vertebrae, which eventually would find their way into the hands of scientists. However, not before a society of Scottish naturalists had declared the Stronsay specimen to be a new and totally unknown creature – most likely the infamous sea serpent of legend. This assessment was based solely on Sherar's sketch, as well as sworn affidavits by the laymen who had studied it. Yet this premature conclusion was soon debunked by accomplished London surgeon and anatomist Sir Everard Home who, had manage to acquire some of the cartilaginous neck bones and immediately recognized them as belonging to a massive, male, basking shark *(Cetorhinus maximus)*. Evidently, the huge fish's advanced state of decomposition had rendered it unidentifiable to those who had encountered it.

You see, when a basking shark dies, the soft gill arches quickly rot away, causing its enormous jaws to fall off and leaving behind a relatively small skull, as well as an extended vertebral column that resembles a long neck. The six "legs" that were noted were no doubt the shark's pectoral, pelvic, and (male) clasper fins in an advanced state of decay. And the fibrous "mane" on its back was likely a combination of rotting dorsal fins combined with the top part of the caudal fin. To this day, vertebrae of the Stronsay Beast are in the custody of the National Museum of Scotland, and Dr. Home's basking shark conclusion has been subsequently reinforced by other academics in residence there.

There does still remain a mystery, though, as the largest basking shark ever scientifically documented was barely forty feet long – and all of the men who had measured the Stronsay carcass were in agreement its length was definitely a good fifteen feet longer than that. Ergo, cryptozoologist and comparative physiologist Dr. Karl

Shuker and I have discussed this discrepancy and share the opinion the Stronsay Beast was probably an outsized specimen of basking shark, perhaps the last of its kind – and in an era before manmade influences such as commercial fishing eradicated these veritable giants of the ocean.

Dr. DeWitt Webb and his St. Augustine Monster (Public Domain)

ST. AUGUSTINE MONSTER

WHILE CYCLING ALONG ANASTASIA BEACH JUST OUTSIDE OF ST. Augustine, Florida, on the evening of November 30th, 1896, two young men noticed a colossal mass of organic matter rotting on the beach. Thinking it significant, the two boys alerted Dr. DeWitt Webb, a local physician and ardent naturalist, who arrived on the scene within days. Webb was summarily astounded by what he saw, estimating the amorphous, unidentifiable blob must weigh at least five tons. He measured its greatest girth at close to twenty feet across. Over the course of the following weeks, the thing would get washed back out to sea on two separate occasions, but ended up back on shore in nearby locations. The good doctor put forth a strong effort to document the anomaly before it was lost for good. Excavating the surrounding sand, he discovered fleshy segments that resembled sections of titanic tentacles and concluded the huge mass must represent the remains of an octopus of literally mind-blowing proportions. Hiring a number of men and horses armed with ropes and plywood, Webb dragged the carcass farther up the beach, took a number of photo plates and (after apparently much effort) managed to hack off some small pieces of flesh before the ocean ultimately reclaimed the odiferous object.

Next, Dr. Webb was able to get in touch with Addison Verrill – at the time the world's leading authority on mollusks who (based on his correspondence with Webb) declared the specimen was most likely a truly monstrous octopus. Amusingly, Webb admitted to Verrill he had spent a veritable fortune on his giant octopus recovery project – a whopping ten dollars! Notwithstanding, when Verrill finally received the tissue samples that Wood had procured, he reversed course and concluded the corpse was probably nothing more than blubber from a sperm whale *(Physeter microcephalus)*. Dr. Webb was evidently unconvinced. The matter remained unresolved for decades.

That is until the 1970s, when a marine biologist named Forrest G. Wood became intrigued by the mystery. After recruiting a rising young octopus specialist named Joseph Gennaro to the cause, Wood obtained one of DeWitt Webb's original tissue samples, still in the collection of The Smithsonian Institution, and the scientists conducted a scientific analysis of the specimen. Unfortunately, the cellular structure wasn't clearly visible under a microscope. But using a polarizing light for improved detail, Gennaro determined the fibrous material contained wide bands of connecting tissue, most similar to those of an octopus. In 1986, biochemist turned Nessie investigator Dr. Roy Mackal undertook his own amino acid study of the tissue and reached the same conclusion as Gennaro. Then, in 1995 biologists from the University of Maryland challenged both findings by subjecting the material to chemical testing and concluded the substance was definitively the protein known as collagen, essentially whale blubber. The entire matter may have been put to rest in 2004 as the result of DNA testing, which verified that a sequenced genome from the flesh of the St. Augustine Monster was indeed identical to a whale.

TRUNKO

ONE TRULY STRANGE STORY INVOLVES A BEACHED CADAVER that is affectionately known within the field of cryptozoology as 'Trunko,' which was likened to a cross between a polar bear and an aquatic elephant, no less! According to newspaper articles published in the early 1920s, the epic episode unfolded at Margate on the coast of South Africa. The principal eyewitness was apparently landowner Hugh Ballance, (whose son would incidentally go on to become a decorated WWII fighter pilot and rat trap inventor!). One November day, Hugh Sr. glanced out at the ocean and beheld a staggering sight – two killer whales about 1,300 yards distant seemed to be engaged in a tussle with a creature that resembled a gargantuan, swimming polar bear! What's even weirder is the shaggy, white beast appeared to possess a long tail that it was using to whip at the attacking orcas! This was evidently all clearly visible to Hugh, as he had managed to retrieve powerful binoculars with which to view the incomprehensible event. The battle apparently continued for some three hours even as Hugh was joined on the beach by a crowd of curious onlookers. Eventually, the whales seemed to have won over and departed. And in the following days, the lifeless body of their victim washed up on the shore, where it was viewed and photographed by the perplexed locals. The animal, if you can call it that, was said to be about forty-seven feet long, covered in ivory fur and with an extended tail attached. Though remarkably, instead of a head, its front end seemed to consist of only a five-foot-long, trunk-like appendage. The aberration rotted on the beach for a period of ten days, but no scientists ever came to examine it. The tide eventually swept the monstrosity back out to sea, never to be seen again.

Unraveling the mystery, cryptozoologists Markus Hemmler and Dr. Karl Shuker concluded Trunko had been in reality nothing more than a heap of whale blubber, similar to the St. Augustine Monster before it – and that its "hairy" white covering, trunk, and tail were nothing more than the stringy remnants of tissue filaments. This assessment was based on three original photos of Trunko that Hemmler and Shuker tracked down in an old, forgotten magazine article. Moreover, it has been suggested the mass never was alive when Hugh Ballance and the other eyewitnesses had observed it offshore. One possibility being the killer whales had merely been jostling the fleshy object about in the surf, giving it an animated quality.

The so-called Santa Cruz Sea Serpent (Public Domain)

SANTA CRUZ SEA SERPENT

IN CHAPTER THREE I MENTIONED TRADITIONS OF A SEA serpent known as The Old Man of Monterey were prevalent along the coast just south of San Francisco, California. Yet the remarkable creature that washed ashore just north of the city of Santa Cruz during the summer of 1925 would ultimately prove to be something else entirely. The animal's lifeless and smelly remains were found on the rocky waterfront by landowner Charles Moore, and word of the odd discovery soon attracted scores of curiosity seekers. About thirty-six feet in length, the beast seemingly possessed a barrel-sized head that resembled that of a duck (complete with bill) attached to a stretched, twenty-foot neck that terminated in a lumpy body with a fishy tail. Perhaps most alarmingly, there appeared to be stumpy, elephant-like legs sprouting out, as well. Several photographs were taken, and at some point, a mysterious, self-proclaimed natural history expert named E. L. Wallace appeared on the site and declared the deceased monster was a prehistoric reptile called a plesiosaur, whose carcass must have been encased in a glacier for eons before thawing out and drifting across the Pacific Ocean to its current location. In the end, the specimen's skull was turned over to legitimate scholars at the California Academy of Sciences who quickly identified it as belonging to a rare Baird's beaked whale *(Berardius bairdii)*. Incontrovertibly, the dead cetacean's startling appearance had been due to the fact that its head had become separated from its body, causing its skin to expand, stretch, and clump in various places like salt water taffy.

GLACIER ISLAND CARCASS

ON NOVEMBER 26TH, 1930, THE *NEW YORK TIMES* PUBLISHED A sensational news article describing how the corpse of a huge, hairy "lizard-like" creature had been found on Alaska's secluded Glacier Island. According to the report, none of the locals could identify the thing, leading some to speculate (like the Santa Cruz Sea Serpent before it) it might be a prehistoric beast that had thawed out after spending countless millennia entombed in glacial ice. The initial description stated the dinosaur-like animal was forty-two feet long, including a head six feet in length, a twenty-foot body and sixteen-foot tail. Its elephantine pate with trunk-like snout spawned suggestions it might be the rotting remains of a woolly mammoth, though its great length, narrow body, and tail seemed to rule out that possibility. A follow-up article by the *Times* a week later reported some forestry workers from the mainland had mounted an on-site investigation and acquired more accurate measurements, though they, too, were stumped as to the true identity of the carcass.

For many decades afterward, this vague and perplexing incident reverberated in cryptozoological literature. That is, until May 02nd, 2008, when an Alaskan journalist named Dixie Lambert penned an article for the *Cordova Times* newspaper that revealed the whole story. The carcass had originally been sighted by a pair of fox farmers (didn't know there was such a thing!) on November 10th, 1930, floating in Eagle Bay. Using a boat, the two men dragged the object onto the shore and hacked off some meat to feed to their foxes. Furthermore, the creature's body had only been about half as long as reported in the original *Times* article. Yet the most important revelation in Lambert's piece was the fact that the skull and skeleton of the animal had been saved and were definitively identified as belonging to a minke whale *(Balaenoptera acutorostrata)*. To this day, the remains are in the collection of the National Museum of Natural History in Washington D.C.

GOUROCK CARCASS

REVERED CRYPTOZOOLOGIST DR. KARL SHUKER ONCE described this mystery beast as representing "one of the most enigmatic sea monster carcasses on record." However, like the Glacier Island Carcass, new details have emerged in recent years that seem to have solved the puzzle. In the early part of June 1942, the remains of the creature in question washed up at Clyde Firth in the town of Gourock, Scotland (and only about one hundred fifty miles south of Loch Ness). A local official named Charles Rankin was put in charge of disposing of the smelly corpse, which was twenty-seven feet in length, with a small, flattened skull attached to a long, tapering neck, a thick body five feet across, four flippers, and long tail. Due to WWII raging at the time – and because the estuary was a strategic location that harbored British Navy vessels – photography was strictly prohibited.

Rankin apparently examined the perished animal, which he later characterized as being relatively intact, with little sign of decomposition. According to his notes, the mouth was full of long, sharp teeth, and the body was covered with pointy bristles. The stomach contents were said to contain part of a sailor's jersey and the tasseled corner of a tablecloth! Rankin removed one of the exterior bristles and then ordered the cadaver be hacked up and buried under what is now the soccer pitch of a local schoolyard. Based on Charles Rankin's description of an "unmarred" body and large teeth, the rotting remains of a basking shark similar to the Stronsay Beast did not seem to be a plausible explanation. That is until 2012, when intrepid investigator Markus Hemmler undertook an extensive inquiry and learned the *Gourock Times* newspaper of the time had run multiple stories describing the monstrosity as "a basking shark in an advanced state of decomposition." At the time, this evidently had also been the conclusion of Dr. A. C. Stephen from the National Museum of Scotland. Still, because basking sharks are filter feeders with tiny teeth, Rankin's portrayal of the Gourock creature as having an unblemished form and carnivorous dentition remains somewhat of a mystery.

TASMANIAN GLOBSTER

OUR NEXT CONFOUNDING CARCASS WAS SO UNUSUAL IT actually inspired a new word – 'globster'! Seemingly like something out of a 1950s science-fiction flick, the team of experts who first examined the thing on the beach described it as a sort of fleshy, flying saucer – pale, rubbery and tough – some twenty feet across and weighing several tons, but covered with fine hair. The creature, if that's what you could even call it, was first discovered by some cattlemen in August of 1960 on the northwest coast of the Australian island of Tasmania. It took over a year and a half for word of the macabre monster to inspire scientists to look into the matter. Then, by all accounts, it took researchers hours to hack off a small piece of the blob, so thick was its skin. And by the time the sample was being shipped off to a lab for analysis, news of the fascinating find had been plastered all across headlines the world over. This happened on March 08th, 1962. Inquisitive readers around the globe waited with bated breath as the anticipation mounted. Zoologists wondered if an entirely new order of deep sea life form was about to be announced.

By March 19th, an official pronouncement by the Australian government declared, woefully, that the globster had been nothing more than a hunk of whale blubber. Early cryptozoologists including Ivan T. Sanderson hinted at a potential cover-up, and for years the Tasmanian oddity remained fodder for paperbacks touting the unexplained. At any rate, a scientific paper published in 1981 confirmed the mass had in truth been compressed fatty tissue – collagen from a long-dead whale. The fine hairs covering it were merely fibers formed by the slow process of decay. In 1997, a second, smaller, Tasmanian globster washed up at a place called Four Mile Beach on the east coast of the island.

TECOLUTLA MONSTER
IN MARCH OF 1969, A GARGANTUAN CARCASS CAME ASHORE near the resort town of Tecolutla on the eastern coast of Mexico. Still reasonably intact, some farmers had dragged the expired animal onto the beach in order to scavenge its ivory. Initial reports stated the monster was over sixty feet long and weighed some thirty tons. In addition, its body was said to be covered in armored plates and wool, to have a fluke-like tail, and most amazingly – one titanic tusk protruding from its head! It wasn't long before the familiar claims of a recently "thawed out" frozen, prehistoric creature were making the rounds. Local officials made the decision to have the malodorous thing forklifted into town so they could take photos and make a grand spectacle out of the event. Today, the monster's skeleton is on display at a local museum. It has been identified as belonging to a baleen, sei whale *(Baleanoptera borealis)*. In retrospect, the specimen's "tusk" had been one of its lower jaw bones that had broken and was protruding outward, and its reported "armor" most likely characteristic skin folds and grooves running from its throat to its belly.

ZUIYO-MARU CARCASS

EASILY THE MOST CELEBRATED AND STILL-CONTROVERSIAL carcass ever dredged up from the sea is the so-called Zuiyo-maru monster, which was hauled up at 10:30 a.m. on April 25th, 1977, by a Japanese fishing trawler, sailing thirty miles east of Christchurch, New Zealand. The odoriferous corpse had been submerged in about nine hundred feet of water when the ship's expansive net pulled it from the depths of the South Pacific Ocean. Instantly recognizing the weird nature of their accidental haul, the man in charge, thirty-nine-year-old production manager Michihiko Yano, ordered the crew to attach ropes to the object and drag it onboard. The men attempted to raise the thing up, but it slid free from its suspension ropes and fell onto the deck with a loud thud. Not wanting to waste the opportunity, Yano decided to give the cadaver a closer look. He also managed to take five photos while its head was temporarily elevated. Over thirty feet long, the fleshy, white and red carcass weighed an estimated two tons and was quite obviously in an advanced state of decay. Possessing a relatively small, flat, turtle-like cranium attached to a lengthy neck and large body, the specimen also seemed to display four massive flippers. Its bottom parts and tail had previously dropped away. "The surface of the body was loose and had white fat," Yano later recalled. "I could see flesh here and there, but it was red and rotten." In fact, the state of decomposition was such that there was a gaping split running down its backside and a bony, vertebral column was clearly visible. Yano guessed the animal had probably been dead for at least a month. Based on its sheer size, his initial thought was the creature must have been a whale, then considered it might be a giant sea turtle missing its carapace. After about an hour of taking measurements and plucking forty-six "horny fibers" from its appendages (as well as letting members of the eighteen-man crew render an opinion), the carcass was dumped back into the ocean, before it could contaminate the haul of mackerel onboard.

When Yano finally returned to Japan over six weeks later, he had the photos developed and turned them over to his superiors at Taiyo Fisheries, along with a sketch of what he had observed. A media sensation ensued when, at a press conference, famed zoologist Yoshinori Imaizumi from Tokyo's National Museum of Science announced the carcass did not match the description of any known living animal, but instead closely resembled a long-necked, prehistoric, marine reptile known as plesiosaur – certainly a scientific discovery of monumental proportions! Imaizumi's conclusion was echoed by fellow Japanese scholars, though other academics from around the globe publicly rejected the possibility. Nevertheless, because prehistoric monsters are obviously revered in Japanese culture, the plesiosaur angle was pushed by newspapers for a solid year. A children's action figure and commemorative postage stamp paying homage to the Zuiyo-maru creature were even released.

Then, in July of 1978, a coalition of Japanese fisheries scientists published a series of nine independent research papers destined to disappoint. Like the Stronsay Beast one hundred seventy years earlier, the carcass had most likely been a rotting basking shark – not a plesiosaur. Several clues were pertinent. First, the number of cervical vertebrae drawn/described by Yano (six or seven) was inconsistent with plesiosaurs, which possessed five times that many. But it was totally consistent with a shark. Also, the short rib lengths measured by Yano again indicated shark over reptile. Additionally, whereas Yano's sketch suggested there was no dorsal fin present, examination of one photo clearly showed a large dorsal fin was in fact present and had simply fallen down along the back, resembling a flap of skin. The clincher was the fact that microscopic and chemical analysis of the fibers Yano had collected could only be from a shark due to the presence of a diagnostic protein known as elastoidin. Still, to this day many publications, TV shows, and websites continue to perpetuate the myth that the Zuiyo-maru carcass is proof that dinosaur-like sea monsters do exist.

Gambo – an unidentified carcass examined by a young naturalist © Bill Rebsamen

GAMBO

AS YOU'VE PROBABLY NOTICED, THE BULK OF OUR BEACHED SEA beasts have generally been found in an advanced state of decomposition, resulting in often false assumptions regarding their true identity. But this is not the case with respect to a largely intact carcass observed by a young naturalist named Owen Burnham, on the morning of June 12th, 1983, at Bungalow Beach in the African nation of Gambia. According to a brief item that appeared in *BBC Wildlife* magazine in May of 1986, Owen was on holiday with his family, strolling down the beach when they came upon two local men attempting to hack off the head of a deceased marine animal. Having lived in the region, Burnham considered himself largely familiar with a variety of local wildlife, but could not identify the creature. Based on his brief examination (including measurements), Owen determined the specimen was at least fifteen long and five feet across; it was black or brown over most of its body, but was white or light-colored underneath; it had a bulbous head and long rostrum (snout) lined with eighty teeth, four paddles, and a long, tapering tail. No fins were present. Sadly, no one in the Burnham family had a camera on hand at the time.

Subsequently, one theory holds that, based on the cryptid's head shape, coloration and dentition, what Burnham saw was a slightly mangled or mutated Shepherd's beaked whale *(Tasmacetus sheperdi),* an incredibly rare, deep sea cetacean. But, its nonfluked tail and apparent presence of nostrils instead of a blowhole argue against that identity. More romantic suggestions include a surviving mosasaur or prehistoric crocodile. It's also possible that a young and overzealous Burnham misinterpreted certain features on the corpse. A 2006 expedition mounted by England's Centre for Fortean Zoology (CFZ) attempted to locate the precise location where the remains had been, hoping to dig up its bones in the sand. But unfortunately a beach club has since been built over the spot. The CFZ team was able to locate a local man who claimed to have witnessed the odd cadaver back in 1983 and stated the thing had merely been a dead dolphin. Although when it comes to describing animals, many people are not familiar with taxonomic differences and tend to use generic names. "Dolphin" is unquestionably a generic term.

BENBECULA BLOB

IN ADDITION TO NESSIE, SCOTLAND CERTAINLY SEEMS TO HAVE its share of mystery carcasses. Case in point: in 1990, a vacationer named Louise Whipps stumbled upon a twelve-foot-long "something" while strolling along the beach on the Hebrides island of Benbecula. Louise posed for a picture with the organic mass, which she described as being fur-covered with something like a head on one end and a tail on the other, although the most notable features were a number of (ten or so) thick lobes, running along its side. Based on those structures, which resemble the ones described on the Tasmanian globsters, it's a fair assumption the blob was yet another mass of whale collagen. Seemingly, these tentacle-like appendages were formed by compressed tissue masses that corresponded with where the rib cage or other body parts of the whale were once situated. Interestingly though, when Louise took her photo of the weird carcass to the Great North Natural History Museum in Hancock, England, the experts there were unwilling to offer a conclusive determination regarding its true identity.

CHILEAN BLOB
THE SCIENTIFIC WORLD BECAME ABUZZ ONCE AGAIN DURING July of 2003, when it was announced the Chilean Navy had discovered an uncanny, gargantuan carcass at Pinuno Beach, near the city of Los Muermos. Almost forty feet across and weighing fourteen tons, the slimy, unidentified mound immediately conjured up images of an immense octopus. Extensive tests conducted on its tissue at the University of Chicago the next year confirmed the mass was merely the adipose (fatty) remains of a long-deceased sperm whale.

PHILIPPINE GLOBSTERS
THE MOST RECENT NEWS FLAP INVOLVING CONFOUNDING carcasses occurred in 2017 and involved a pair of amorphous, shaggy, white masses that washed up in the Philippines Islands. The first was twenty feet long and weighed over four thousand pounds. It came ashore on Dinagat Island in the south during the month of February. The second was discovered farther north on Oriental Mindoro on May 11th. It was over twice as big as the first one at around fifty feet long. Pictures and videos of both

extraordinary organisms lit up social media. Marine biologists confirmed both had been the remains of decomposing sperm whales and the hair-like substance on their exteriors was merely fibrous, shredded muscle tissue. It is now believed the animals perished as the result of New Zealand's Kaikoura earthquake of November 14th, 2016. The 7.8 magnitude event reverberated through the surrounding ocean, disrupting the whales' food supply and behavior patterns – ultimately displacing the creatures and causing them insurmountable distress.

WHILE ALL OF THE NOTABLE CARCASSES PRESENTED IN THIS chapter have mostly proven to be decomposing basking sharks and whales, there is still hope that – if there truly are sizable, undiscovered species inhabiting the Earth's oceans, lakes, and rivers – their remains might surface one day.

9

Deep Sea Discoveries & Monster Mimics

'The sea, once it casts its spell, holds one in its net of wonder forever." – JACQUES COUSTEAU

There's no doubt cryptozoologists are eternal optimists. Still, one of the primary reasons the field has been able to persevere is that occasionally a surprising discovery results in some substantial new species being added to the annals of the current zoological order. While many hidden animals are found in the dense, biodiverse jungles of South America's Amazon Basin, the African Congo, or the subtropical forests of Southeast Asia, even the most cautious scientists will acknowledge there are likely large, unknown creatures inhabiting the Earth's oceans. We must remember that 71 percent of our planet's surface is covered with water, most of it incredibly deep – about twelve thousand feet on average. As you might expect, every so often the sea surrenders once of its murky secrets.

Greatly exaggerated giant squid attacks a sailing ship (Public Domain)

GIANT SQUID – RELEASE THE KRAKEN!

THE LARGEST INVERTEBRATE IN THE WORLD WAS COMPLETELY unfamiliar to science until the latter part of the nineteenth century. And there exists a strong parallel to our sea serpents and lake monsters like Nessie since the giant squid *(Architeuthis dux)* was for many years strictly relegated to the world of mythology. Hints of its existence date back to the Ancient Greeks – including the great poet Homer, who wrote about multiheaded sea monsters including the Scylla and Hydra. The multiple "heads" may have been a reference to the many tentacles characteristic of cephalopods such as the octopus and squid. The Scylla in fact was said to possess rows of teeth along its snake-like "neck," an obvious reference to the razor sharp teeth that line the suckers on the tentacles of squids. The

Greek polymath Aristotle, perhaps the very first zoologist, described in writings the now familiar squids found in the Mediterranean Sea, but also referenced a giant species, which he referred to as "Teuthos." Yet the first mention of the infamous monster known as the Kraken came from the Swedish cartographer Olaus Magnus, who first depicted sea serpents and other terrible beasts roaming the world's oceans on his maritime maps. Magnus portrayed the Kraken as a diabolical entity that lurked off the coast of Norway – one that resembled an uprooted tree. Two hundred years later, the Norwegian naturalist Erik Pontoppidan mentioned a multi-armed monster a mile and a half in circumference (an obvious embellishment), one capable of sinking ships by way of the massive whirlpool it left in its wake. The Kraken was also known as the "Island Fish" because mariners would often mistake it for a landmass, only discovering its true nature once they had climbed onto its surface, at which point the colossal creature would sink into the depths, dragging its victims downward to watery graves. This, of course, is another sensational exaggeration. But the underlying idea there was a humongous, hidden beast beneath the waves persisted.

Then, in the early nineteenth century, a French naturalist named Pierre Denys de Montfort began putting it all together. The scholar published a comprehensive volume on mollusks and suggested there were both monster-sized octopuses and squids, fully capable of taking down vessels, or at least plucking sailors off of ships' decks with their massive tentacles. This, too, might have been a bit of a stretch, though it should be noted there are documented cases of giant squids that surfaced and pulled soldiers off of life rafts during WWII. To his credit, de Montfort also provided examples of incidents where the carcasses of giant squids had been found beached in the past, including one in Iceland in 1639. The zoological establishment still considered it all purely fiction. Picking up where de Montfort had left off, a Danish naturalist named Japetus Steenstrup penned a scientific paper in 1857, officially describing the giant squid as a new species. Steenstrup had even managed to obtain a large, parrot-like beak from one of the creatures, which had been stranded and perished on the Norwegian coast. A pivotal incident transpired when a French corvette (small warship) called the *Alecton* sighted a giant squid floating near the Spanish Canary Islands. After pummeling the poor thing with cannonballs, the crew almost managed to drag the slimy corpse onboard before the squid's lower half broke off and sank back into the depths. The captain begrudgingly ordered his men to discard the top half, too. Finally, in 1873 a squid that had attacked a small fishing craft off of the coast of Newfoundland, Canada had most of its thirty-foot tentacle hacked off with an axe. Marine biologists were forced to accept the fact that the Kraken was now a reality.

THROUGHOUT THE 1870S, MASS STRANDINGS OF GIANT SQUIDS on Newfoundland and elsewhere introduced another element of mystery. Scientists still really aren't sure why these rare events occur. But it appears as though every few decades, large numbers of these typically elusive, deep sea giants are found floating dead or washed up on beaches around the world. Still, despite their great size – with scientifically documented lengths of forty-three feet and weights approaching half a ton, very little is known about these colossal calamari, which thrive thousands of feet below the surface and are rarely seen. It wasn't until 2002, well over a century since its discovery, that a live giant squid was filmed in its natural habitat. And on June 18th, 2019, just the third-ever video of a giant squid was obtained by a camera placed at great depth, about one hundred fifty miles off the coast of Louisiana in the Gulf of Mexico. Summarily, we must ask the question: If these genuine monsters of the deep could remain hidden for so long, what else could be out there?

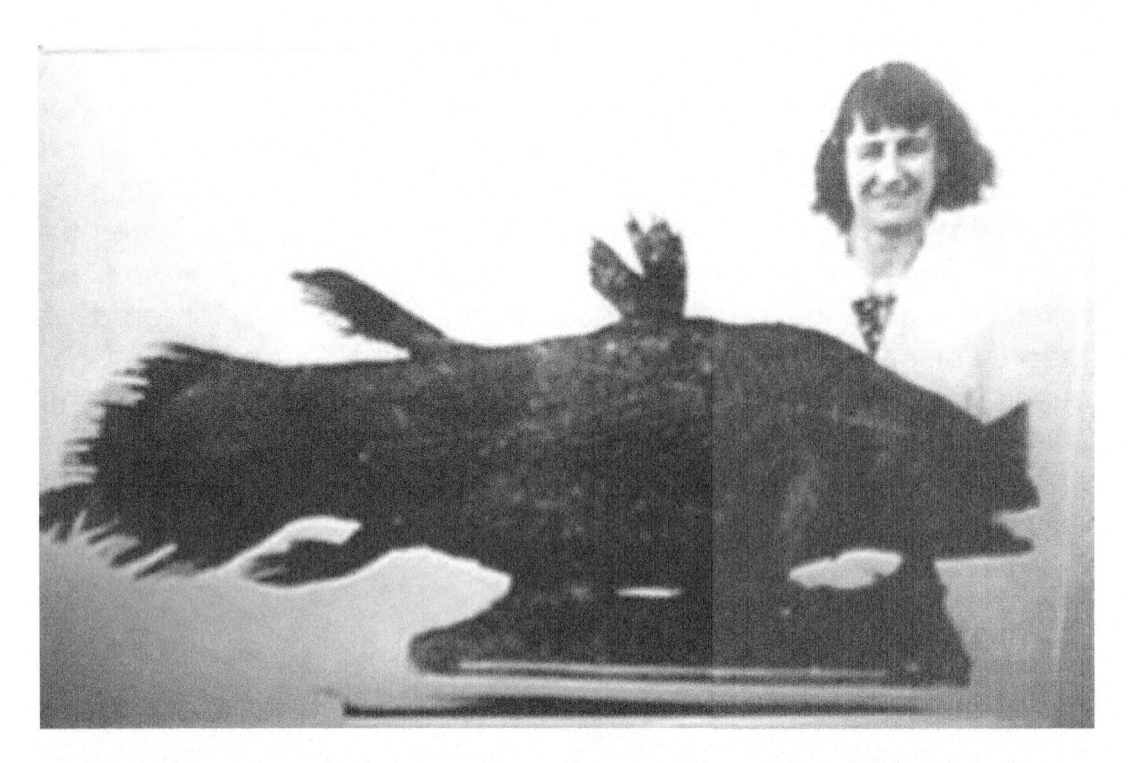

Marjorie Courtenay-Latimer with her mounted coelacanth (Public Domain)

COELACANTH

ARGUABLY, ONE OF THE GREATEST ZOOLOGICAL DISCOVERIES OF the past century features the coelacanth *(Latimeria chalumnae)*, a large, idiosyncratic fish, which belongs to an order that scientists thought had died out with the dinosaurs, sixty-five million years ago. The saga began on the morning of December 23rd, 1938, when a trawler skippered by Captain Hendrik Goosen was dragging its nets about three miles offshore near the mouth of the Chalumna River, on the southern coast of South Africa. When the vessel returned to port with its impressive haul, it was met at the dock by Marjorie Courtenay-Latimer, curator at the nearby East London Museum. Marjorie often checked her friend Captain Goosen's catch, in order to see if there was anything of interest to the museum. On this particular day there most certainly was! Sorting through the standard fish one at a time, Marjorie was amazed to see a shiny, turquoise fin that she didn't recognize sticking out from the bottom of the pile. When the curious scholar was at last able to completely clear it off, Marjorie was mesmerized by the sight of a bulky, five-foot-long fish covered in beautiful, azure, armor-like scales. The animal also displayed "large blue eyes," an unusual, half-lozenge-shaped tail, two dorsal fins, and (most remarkably) four "limb-like fins" on its underside. She had absolutely no idea what the thing could be.

Fortunately, Marjorie had the wherewithal to take the trophy fish back to her museum and make a sketch of its unusual features. Though sadly, she had no way to slow down the decomposition process, so she took the specimen to a taxidermist in order to have it cleaned out and mounted. Next, Courtenay-Latimer mailed her sketch to a friend and colleague, Dr. J. L. B. Smith, a trained chemist who was also a skilled ichthyologist and avid fisherman. Because he was on holiday at the time, Smith didn't receive Marjorie's sketch until ten days later. But when he first gazed upon the crude drawing, he literally couldn't believe what he was seeing. Smith calmly turned to his wife and declared, "Very soon, the name of this fish will be on the lips of every scientist in the world." In February of 1939, when he was finally able to visit Marjorie at the museum, Smith told her finding a living coelacanth was akin to walking down the street and running into a dinosaur. He subsequently named the genus *Latimeria* in its discoverer's honor.

Of the utmost importance to the now obsessed scientist was what knowledge the skeletal structure and internal organs of the coelacanth might reveal. So Smith began a fourteen-year quest to obtain the carcass of a specimen that was still wholly intact. Speculating the first one had merely been a stray that'd been driven south by the currents, he concentrated his search up the coast of East Africa, including Madagascar and the Comoros Islands – distributing thousands of multilingual flyers that depicted the coelacanth and offering a substantial reward for one that had been freshly caught. The scientist's efforts finally paid off in December of

1952, when Smith managed to obtain his coveted prize, reeled in by a fisherman off the Comoros Islands, just as he had predicted. In order to ensure he could preserve the specimen in its entirety, Smith convinced the president of South Africa at the time to enlist an Air Force plane in order to get him to the Comoros Islands and back to his lab with his treasure as quickly as possible.

Comprehensive studies of the coelacanth's unique anatomy through the years have in fact revealed a number of peculiar and ancient features. Instead of a backbone, the fish possesses a stiff, stabilizing rod known as a notochord. Also, the species displays several shark-like characteristics: jelly-filled electroreceptors on its snout in order to detect prey and a swim bladder filled with oil instead of air. It is also ovoviviparous (giving birth to live young). In addition, the coelacanth retains an exceptionally primitive heart for a vertebrate. But most remarkably, its rigid, lobe-like limbs give scientists a clue as to what the first fish to crawl out of the sea some four hundred million years ago might have looked like. It is these notable appendages that have earned the pioneering coelacanth the nickname "Old Fourlegs." In this respect, some view the fish as a sort of "missing link" between aquatic and amphibious/terrestrial animals. If nothing else the coelacanth remains, as author Loren Coleman calls it, "The Darling of Cryptozoology," by demonstrating the kind of surprises the sea may bring forth, as well as the role of serendipity in scientific discovery.

MEGAMOUTH SHARK

ON NOVEMBER 15TH, 1976, A US NAVY RESEARCH VESSEL WAS operating just off the northern coast of the Hawaiian Island of Oahu. The ship had deployed two huge, umbrella-type anchors to a depth of five hundred feet. When the moorings were brought up, its crew was astonished to see a monstrous shark tangled in one of the rigs. Sadly, the animal quickly perished. But because all on board assumed it was a juvenile whale shark *(Rhincodon typus)*, it was decided to take the remains to the Navy's lab at Kane'hoe Bay in case it was of any value. Meanwhile when young marine biologist Leighton Taylor was summoned to the scene by the lieutenant on duty, he was perplexed. The animal wasn't a whale shark, and he couldn't identify the species. Taylor had the fourteen-and-a-half-foot, fifteen-hundred-pound specimen loaded into a truck and drove it back to the Waikiki Aquarium where he was director. After placing the cadaver into cold storage, Taylor then called one of his experienced college professors, ichthyologist Dick Rosenblatt, and described the shark as best he could. To which Rosenblatt answered, "I've never seen or heard of anything like that."

As it turns out, the megamouth shark *(Megachasma pelagios)* was not only a new species totally unknown to science, it was so unique that it represented an entirely new genus as well! Named for its cavernous maw, the megamouth is similar to both the whale shark and basking shark in that it is a filter feeder, swimming slowly around with its jaws wide open – scooping up clouds of tiny crustaceans called zooplankton. The edges of the creature's gaping, tiny-toothed muzzle even sport bioluminescent 'lips' that lure in curious prey. Over four decades later, there have only been about sixty-two megamouth sharks in total documented around the world, making it an incredibly rare creature, despite its impressive size. In fact, it appears there was absolutely no ethno-knowledge of this large animal whatsoever – no legends or anecdotes in the Pacific Island culture that would even hint to its existence. It is believed these enigmatic fish spend most of their time at great depths, only coming near the surface on rare occasions – similar to the giant squid, the coelacanth, and perhaps also to our hypothetical sea serpent.

BEAKED WHALES

ON A WINDY, OVERCAST EVENING IN JUNE 2014, BIOLOGY teacher and avid birdwatcher Christian Hagenlocher was scanning the beach of Alaska's desolate St. George Island, hundreds of miles off the Alaskan mainland. Suddenly, he noticed the carcass of what appeared to be a small whale sprawled out on the jet-black volcanic sand at the edge of the surf. He'd visited the very same beach earlier in the day and the carcass hadn't been there at that time. Christian initially thought the animal might be a dead beluga and felt that he should report the find. He promptly alerted his friend marine biologist Karin Holser, who lived on the island. Upon examining the cadaver with Christian, Holser remarked it resembled a Baird's beaked whale *(Berardius bairdii),* a rarely seen cetacean, although this specimen appeared to be only about two thirds the typical size (only twenty-four feet long), with darker skin and well-worn teeth that seemed to be those of a full-grown adult. A tissue sample was ultimately taken, and when a DNA test was conducted, the St. George whale was found to be an entirely new species. This launched a comprehensive study, which determined that other, similar, whale remains had previously been collected, but not properly classified. Remarkably, a matching DNA sequence was found to belong to a whale skeleton that was hanging in a high school auditorium in the Aleutian Islands! That deceased animal had washed up in 2004. Another matching sample in Japan had come from a 2013 carcass that had turned up there. As it turns out, Japanese whalers had long told of a small, jet-black, 'Karasu' (raven) whale, yet science seemed to be lagging behind as far as putting the pieces together. Finally, in September of 2019 a team of Japanese scientists made it official – officially describing the smallest type of Baird's beaked whale *(Berardius minimus)* and adding it to the taxonomic system of known species.

    The beaked whales (Family: Ziphiidae) you see are among the most mysterious large animals in the world. There are twenty-two known species so far, many of them only discovered in recent decades. For example, the pygmy beaked whale *(Mesoplodon peruvianus)* was only officially classified in 1991. Generally speaking, beaked whales are among the smallest of the cetaceans, ranging from thirteen to forty-two feet in length and weighing less than one and a half tons. In form, they look similar to porpoises due to their long beaks, though almost all of the known species are essentially toothless. These highly elusive creatures are virtually never seen due to the fact that they live far out at sea and spend most of their time at great depths – thousands of fathoms deep in some cases. In 2017, the longest recorded dive by a whale was recorded when a Cuvier's beaked whale *(Ziphius cavirostis)* that had been tagged by marine biologists spent three hours and forty-two minutes below the surface before coming up for a breath. If our sea serpents and lake monsters are snake-like whales descended from the ancient archaeocetes, they may display similar behaviors to beaked whales, explaining why we have not yet found conclusive evidence of their existence.

KEEPING THINGS IN PERSPECTIVE, WE MUST ALSO acknowledge that without a doubt there are known species that may have been mistaken for sea serpents and lake monsters from time to time, on account of their great size and highly unusual appearance. Here I will list a few of the leading candidates.

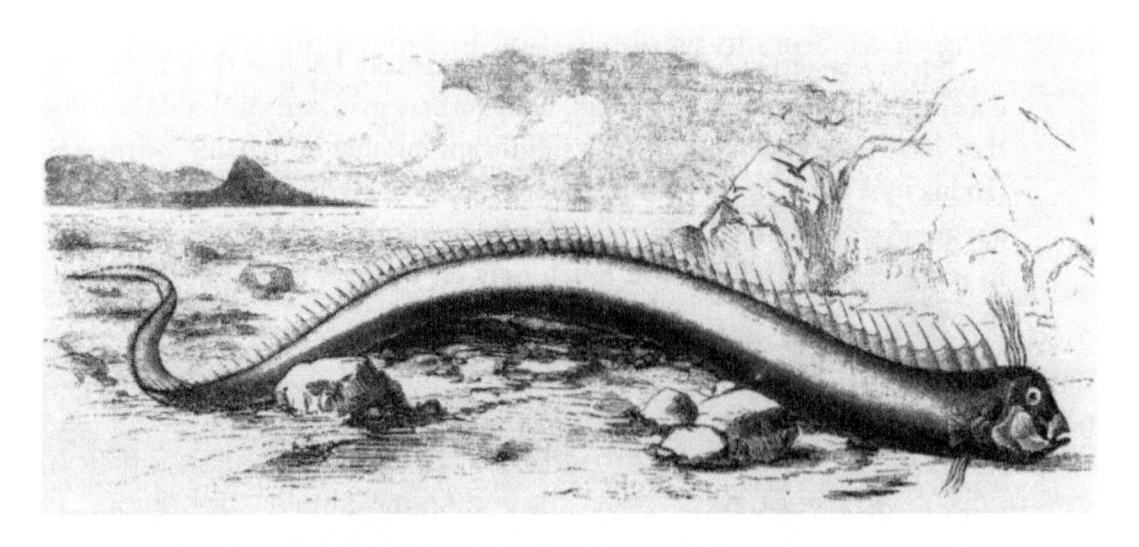

Rendition of an oarfish that washed ashore in Bermuda in 1860 (Public Domain)

OARFISH

I'VE ALWAYS CONSIDERED MYSELF AN EXTREMELY LUCKY angler. Granted, I haven't fished nearly as much as many of my close friends have through the years, and honestly it's been awhile. But, when I was younger I seemed to do exceptionally well for myself. Growing up in Minnesota, I once pulled an impressive largemouth bass *(Micropterus salmoides)* out of such a tiny lake that no one believed I'd actually caught it there. Then, as a young man I reeled in a trophy, twenty-three-pound northern pike *(Esox luscious)* from Reindeer Lake in Northern Saskatchewan. The resulting mount still hangs proudly on my wall. Later, when I moved to Texas, I remember casting just one time off of a bridge into Freeport's Intracoastal Waterway and bagging a nice size flounder *(Paralicthys lethostigma)*. But without a doubt, the most startling fish I ever caught was off of the Galveston Jetties one summer evening. When I reeled in the line, I literally couldn't believe what I was looking at. The thing was about a foot-long, flat as a pancake and shiny silver, but possessed a spiny Mohawk running the length of its snake-like body, as well as huge eyes and a mouth full of razor sharp teeth. In truth, it looked like a baby sea serpent. Several other anglers came over to gawk at the queer catch, though no one could identify it. Later, I discovered I had hauled in a juvenile Atlantic cutlassfish *(Trichiuris lepturus)*, often referred to a ribbonfish.

Now, there is a different type of so-called ribbonfish found throughout the world's oceans and that, due to its great size, may be partially responsible for inspiring elements of the sea serpent legends. Also known as the giant oarfish *(Regalecus glesne)*, due to the fact that its body is uniformly compressed and flat like an oar, these weird, elongate creatures can grow up to twenty-six feet in length (there are unsubstantiated tales about them growing over fifty feet). Very little is known about their behavior, as they live deep in the ocean and are so rarely seen. In addition to their unusual shape, giant oarfish are metallic silver in color, lack scales, display a square, blunt head, and perhaps most startlingly, four hundred scarlet dorsal fin-rays running the length of their back, forming a crest. The six front-most rays sit exceptionally high, like antennae or a rooster's comb. Oarfish are truly peculiar-looking animals.

On January 22nd, 1860, a giant oarfish in the throes of death was found washed upon some rocks on the shore of Bermuda's Hungry Bay, and was at first labeled as a sea serpent. Similar strandings occur every now and then, which is typically the only time that these elusive fish are encountered by man. Marine biologists aren't really sure why this happens, though they speculate the species might be forced into shallow waters by strong storm currents. Japanese folklore states these beachings are warnings of impending earthquakes – and there may be some truth to that.

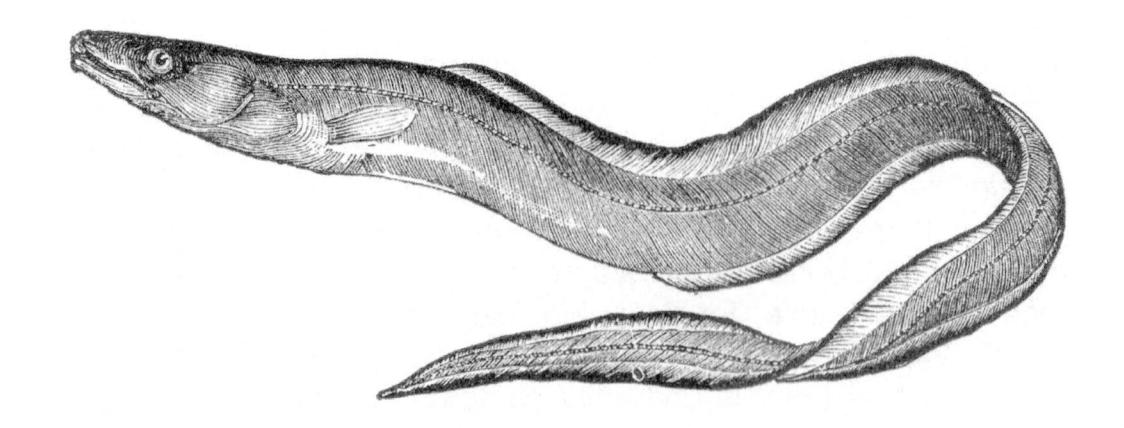

Vintage depiction of a conger eel (Public Domain)

ENORMOUS EELS

FROM THE VERY BEGINNING OF THE LOCH NESS MONSTER ERA, several of the more measured investigators and academics who have taken an interest in the mystery have proposed that Nessie, as well as other lake monsters and sea serpents, represent immense eels. These exceptional specimens are reasoned to be mutants of a sort – either individuals that are born with a genetic condition known as gigantism, which causes them to grow to freakishly massive sizes, or, as cryptozoologist Dr. Roy Mackal first suggested, so-called "eunuchs" – individuals who lose their migratory sex drive and end up living relatively sedentary lives gorging themselves on food and gradually growing bigger and bigger. Although this is strictly a hypothesis at this point.

Eels are elongate fish belonging to the order *Anguilliformes* that are generally very flexible and have smooth, nonscaly skin. They possess a long caudal fin that fuses with the dorsal fin and uniformly wraps around their body, but no ventral fins. Eels tend to live in holes or crevices and venture out (often at night) swimming with horizontal undulations as they search for food such as smaller fish, as well as crustaceans (they can even be cannibalistic!). While there are some eight hundred recognized species around the world, some as small as a few inches in length, the green moray eels of Caribbean waters can grow up to ten feet long. However, the largest species endemic to Loch Ness is the European conger eel *(Conger conger),* which are gray or black in color and average about five feet in length. The biggest congers can be found in the Atlantic Ocean, where they've been known to grow to seven feet and can weigh over one hundred fifty pounds. So when we consider the Loch Ness Monster has typically been assigned a length of at least fifteen feet and is frequently estimated to be as much as forty feet long, it would require a conger eel of truly Biblical proportions in order to explain the sightings.

There are other issues with the eel hypothesis as well. When we consider the thousands of Nessie reports on record, the vast majority of observers have noted a huge, wide hump, like an upturned boat, that sticks a few feet above the surface of the water. Less frequently, witnesses mention a series of vertically undulating humps, or a small head attached to a long neck, which is held aloft. It just doesn't jibe with the physiology of an eel in the slightest, though an eel writhing on its side at the surface might conceivably create the illusion of undulating humps. Still and all, it's an inescapable fact that some Nessie deponents have declared unequivocally that what they saw was indeed an enormous eel. As we discussed in Chapter Two, a recent environmental DNA study of Loch Ness's waters found a surprising abundance of eel DNA, which does not refute the giant eel theory. At the same time, however, many of our other alleged 'monster' lakes around the world are not known to sustain eel populations of any kind.

Beluga sturgeon caught in Russia's Volga River in 1921 (Public Domain)

STUPENDOUS STURGEON

INEVITABLY, A PERCENTAGE OF LAKE MONSTER REPORTS MAY BE the result of misidentifications of known but seldom seen outsized fish – often representing frightening-looking species. From Loch Ness, to Lake Okanagan, to Lake Champlain, to the lesser-known but puzzling case of Alaska's Lake Iliamna, the most commonly suggested candidates includes various species of sturgeon. Perhaps most known for their coveted eggs known as caviar, which can fetch prices close to one thousand dollars an ounce, these ancient, prehistoric-looking fish possess a cartilaginous skeleton like a shark, a bulbous rostrum (snout) with barbells and a string of bony plates called scutes along their backs. They are generally bottom-feeders (called benthic), rarely spotted near the surface and similar to salmon and eels in that they are also anadromous – meaning they regularly travel from the ocean into freshwater rivers and lakes in order to spawn. The largest known species are the beluga or great sturgeon *(Huso huso)* of Eastern Europe, which can live up to a century. One specimen caught in Russia's Volga River in 1827 was reportedly twenty-four feet long and weighed close to thirty-five-hundred pounds! Sadly, due in large part to caviar harvesting, this particular species is now critically endangered.

In North America, the most sizable species is the white sturgeon *(Acipenser transmontanus),* whose range conveniently extends across that of our northern monster lakes – from the Pacific Northwest to the Great Lakes. The largest documented specimen ever was caught in Idaho's Snake River back in 1918. It was twelve feet long and weighed fifteen hundred pounds. However, due to overfishing, truly gargantuan individuals are becoming fairly scarce. In 1987, a dead eleven-foot individual was found floating in Lake Washington, near Seattle. And in 2016, a ten-foot sturgeon known locally as "Pig Nose" (it had lost the tip of its snout in some kind of accident) was captured and then released in British Columbia's Fraser River. However, the record white sturgeon from Eastern North America, caught by a fisherman in Wisconsin in 2012, was only seven feet long and weighed less than three hundred pounds. With regard to Loch Ness, it's worth noting sturgeon are incredibly rare in British waters. So much so that any sturgeon caught in the United Kingdom must be presented to the monarchy! Furthermore, no sign of sturgeon DNA was found in the Loch Ness eDNA study of 2019, so there's little chance these huge fish are Nessie candidates. Ultimately, though sturgeon are certainly capable of reaching the sizes reported in some lake monster accounts, their physical characteristics simply do not match with the majority of eyewitness descriptions.

The author at Texas' Lake of the Pines, said to be home to a monster, car-sized catfish

COLOSSAL CATFISH

AS I TRAVEL AROUND THE UNITED STATES, I OFTEN HEAR A recurring urban legend. It always involves someone who knows a diver that was working underwater on a dam when he encountered a catfish as big as a Volkswagen! Said diver then makes the prudent decision and swims to the surface like Michael Phelps, swearing he will never go down again. I've heard this particular tale related to lakes from places like Arizona and Texas to Missouri and Georgia. Wherever the story stems from, it is interesting how it has been incorporated into different cultures around the country. Nevertheless, there truly are some humongous catfish found in rivers and lakes around the globe. The piraiba catfish (*Brachyplatystoma filamentosum*) of South America's Amazon Basin reaches six feet in length and obtains weights of over three hundred pounds. The European wels catfish *(Siluris glanis)* can grow over seven feet long and can weigh some four hundred. And the largest species in the world, the Mekong catfish *(Pangasionodon gigas)* of southeast Asia – up to nine feet in length and a whopping six hundred pounds! In North America, the largest species is the common blue catfish *(Ictalurus furcatus)*, which by comparison only weighs in at about one hundred twenty pounds, tops. Be that as it may, catfish can live a long time, eat pretty much anything, and grow throughout the course of their lives. Because they prefer lurking near the bottom, on the rare occasions when they do surface, they might be perceived as being monsters.

With regard to Loch Ness, the most likely Nessie candidate would be the wels catfish, though Scotland lies out of the species' accepted range, despite the fact that they can be found virtually all over the rest of Europe. Yet, one theory holds that nobility may have introduced some of these massive fish into Loch Ness generations ago for sport. However, as with our other candidates, the physiology of even a megalithic catfish would hardly explain the descriptions of Nessie. Moreover, the comprehensive 2019 eDNA study at Loch Ness failed to produce any evidence of catfish DNA whatsoever.

DO THE LOCH NESS MONSTER AND SIMILAR CREATURES EXIST? Only time will tell, I suppose. But it's hard to ignore the body of persistent sightings, sonar contacts and photos that have accumulated over the years. It is my ardent hope that within these pages I've given you a glimpse into the abyss of possibilities. The truth is often stranger than we can imagine.

Printed in Great Britain
by Amazon